# THE GATHERING STREAM

JAMES MILLER

# THE
# GATHERING STREAM

## THE STORY OF THE MORAY FIRTH

BIRLINN

First published in 2012 by
Birlinn Limited
West Newington House
10 Newington Road
Edinburgh
EH9 1QS


www.birlinn.co.uk

ISBN: 978 1 78027 095 1


British Library Cataloguing-in-Publication Data
A catalogue record for this book is available from the British Library


Typeset by Brinnoven, Livingston
Printed and bound by Gutenberg Press, Malta

# CONTENTS

# LIST OF ILLUSTRATIONS

# LIST OF COLOUR PLATES

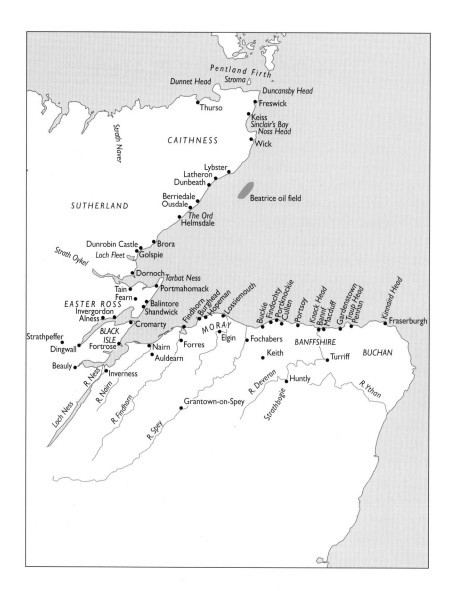

The Moray Firth.

# INTRODUCTION

## Introduction

On a clear day from the Ord of Caithness one can see a thin blue strip of land lying along the southern horizon. This far shore on the other side of the Moray Firth from my birthplace intrigued me long before I had the opportunity to visit it. In time I came to know that, although there are important differences, the two sides of the Firth have been intimately bound to each other and have a shared past. One glance at the map shows that the Firth has the shape of a triangle, cutting into the body of Scotland like an arrowhead thrusting south-westwards. Two sides of the arrowhead delineate about 150 miles of distance but, thanks to the complexity of the inner firths where land and sea meet like interlocking fingers, the actual coastline is much longer. The Dornoch Firth, the Cromarty Firth and the combined length of the Inverness and Beauly Firths define the hammer-headed peninsula of Easter Ross and the leaf-shaped Black Isle. Adding their shores and the basin of Loch Fleet to the length of coastline just about doubles the total. The triangular shape also means that the Moray–Buchan coast is much closer to Caithness and Sutherland than the land-bound traveller might think. Only 30 miles separates Findhorn from Helmsdale, and less than 60 lie between Banff and Wick, whereas the road distances for the same journeys are 98 and 174 miles. This proximity has had important effects in the past.

On land the boundaries of the Moray Firth region are harder to define. The southern border of what was regarded in the early Middle Ages as the province of Moray was the Mounth, the great mass of the Monadhliath and the Cairngorms stretching east from the Great Glen towards the coast of Aberdeenshire. The name of the Mounth is very old and is Celtic in origin; in his book on the Celtic place names of Scotland, W.J. Watson lists its

1

From Burghead there is a clear view of the hills of Sutherland and Caithness, 30–40 miles away. The sliced triangle of Morven, the highest hill in Caithness, can be easily distinguished.

cognates – among them *monid* in Old Breton, *mynydd* in Welsh, *monadh* in Gaelic. Another chain of hills, the spine of the country, what Adamnan in his life of Saint Columba called *dorsum Britanniae*, the watershed between rivers draining to the North Sea and the Atlantic, runs north–south through Ross and Sutherland and forms a western boundary. The landward bounds of the region are roughly defined by these mountain ranges but the space of our story does not always extend so far. It is better simply to go where the history takes us, whether it is to the confines of a single bay or the breadth of the North Sea, and not impose rigid boundaries at the outset.

Tradition has it that James IV used to boast of a town in his realm so long that the people at one end could not understand the tongue of their neighbours at the other. He had Nairn in mind, and the joke referred of course to the meeting of Scots and Gaelic speakers. The language frontier was real, of long standing and the vestige of a more complicated

2

conjunction. The Moray Firth coast was a meeting place for five cultures, each linguistically distinct – Pictish, Gaelic, Norse, Anglo-Saxon and Norman/Flemish. This meeting played itself out over several centuries, reached a peak of confrontation almost a millennium ago and has left its mark. Curiosity about this historical conjunction led to this book and to the metaphor of my title, the in-gathering of influences that change a place as time flows on. The history of the Firth is to a large measure the story of how the people have responded to impacts from outside.

There have been many histories of Scotland and of the counties and towns within the nation. In recent years there has been a welcome spate of books on the history of the Highlands, defined often in opposition to the history of Scotland as a whole. Here I am attempting a slightly different project, a history of a region, spanning several counties but having as a unifying idea a share of the Moray Firth coastline. I am not the first to venture along this path. Cuthbert Graham wrote a general book on the whole region in 1977,[1] and Ian Mowat has studied in detail the impact of the period of great changes between 1750 and 1850 on Easter Ross and the conjunction of Lowland and Highland societies.[2] Most recently, David Worthington has approached the study of the Firth as a particular maritime region.[3]

Over the years I have been fortunate enough to be able to write on several historical themes. In a very real sense, we can never know what the past was like at any particular time for the people who experienced it as their day-to-day reality. Whenever we write history we are imagining and reinventing the past. Facts remain inviolate – the Battle of Culloden was fought on 16 April 1746 and that's that – but a fact is often a meagre piece of knowledge that tells us nothing about the reality of the occasion for the people who were there. Nevertheless I hope that I have created a narrative covering the last 2,000 years that, despite a necessarily broad approach and some speculation, links some of the facts and explains how we have reached where we are today, how the past 2,000 years have created a context for our lives.

One of the first men to attempt a history of Moray was Lachlan Shaw. A farmer's son, born between 1685 and 1690, he studied at the universities of Aberdeen and Edinburgh before becoming the parish minister of Kingussie in 1716 and finally of Elgin in 1734 where he stayed until his death in 1777.[4] 'Before the reign of Malcolm Canmore,' he wrote, 'all is darkness in the history of Scotland at large . . .' This is no longer the case, thanks to the

A testament to the fertility and mildness of lowland Moray as well as to the people's skills and enthusiasm, this noticeboard in Forres records the town's repeated success in the Britain in Bloom contest. Sheltered from the prevailing westerlies, the area has drier, warmer summers than its latitude would suggest.

labours of many researchers in history and archaeology. Another complaint came from a later minister of Elgin, John Grant in 1793, who noted that 'Our historians . . . chiefly employ themselves in retailing legendary stories or giving inaccurate accounts of foreign or domestic wars, and political contests, overlooking unfortunately the more important details of industry, trade and population'.[5] That is a warning that I have tried to bear in mind. It was coincidence that I began to explore the ideas in this book before I learned of the reassessment of the Pictish province of Fortriu as Moray. I hope I have done justice to some of these new ideas.

Many people have given generously of their time and knowledge during the research for this book. I am especially grateful to Dr David Worthington, director of the Centre for History, University of the Highlands and Islands, for discussing with me some of my ideas, and to him and his colleagues for welcoming me to use the Centre's library in Dornoch. Dr Karen Cullen kindly gave me the benefit of her research on the famine years in the 1690s.

The rugged cliffs near Pennan.

In Elgin, David Addison, then the curator of the Elgin Museum, shared his knowledge of the Romans in the north. Dr Alistair Rennie of the Earth Science Group of Scottish Natural Heritage brought me up to date with my understanding of coastal changes. Dr Jim Wilson of the University of Edinburgh provided information about studies on DNA in the people of Scotland. Eric May, secretary of the Friends of the Falconer Museum in Forres; Alasdair Cameron, a local historian on the Black Isle; and Gordon Cameron, curator of the Applecross Heritage Centre, were generous in answering queries and sharing an enthusiasm for the subject. As always, the staffs of the various public libraries around the Firth, especially in Banff, Elgin, Forres, Inverness, Dingwall and Wick, and the staffs in the Highland Archive Centre and the Scottish Natural Heritage Library in Inverness, were a tremendous help in locating sources. Special thanks are due to Juliet Gayton, Jamie Gaukroger of Highland Libraries, Harry Gray of the Wick Society, Avril Hill, Dr Fraser Hunter, Lesley Junor of High

The coast of the Firth from Rosehearty to Pennan.

Life Highland, Dr Doreen Waugh, and Kathryn Logan and Melanie Poole of the Moray Firth Partnership, for their generous help with illustrations. The sources of the illustrations are included in the individual captions. Sinclair Dunnett read the typescript and made many valuable suggestions for its improvement. As ever, I am deeply appreciative of the patience and help shown by my agent Duncan McAra, and by Hugh Andrew, Andrew Simmons and their colleagues at Birlinn. None of the above is responsible for any errors I have made or opinions I have expressed.

James Miller
Inverness
January 2012

# CHAPTER 1
# LONG REACH OF EMPIRE

The ships moved slowly, keeping a safe distance from the unfamiliar shore. Never before had the Roman fleet, the *classis Britannica*, penetrated so far north and no one aboard, with the possible exception of any guides, willing or coerced, knew what to expect. The military governor of Britannia, Julius Agricola, had just defeated the Caledonians in a pitched battle at a place called Mons Graupius and now his admiral had been despatched to explore the extremities of this island on the fringe of the known world. The record of the voyage is brief. It was written by Agricola's son-in-law, Cornelius Tacitus. He relates simply that the sailors enjoyed good weather and returned without loss to their base at Trucculensis Portus, after finding in the north a 'huge and shapeless tract of country', subjugating [*sic*] the Orkney islands and sighting Thule.[1] How far the fleet went remains the subject of debate, but in the Moray Firth at least we can imagine the ships probing along the coast. At the heart of the squadrons were the biremes, the standard battle wagons, 100 feet long, sleek hulls brightly painted or left black from the pitch-treated timber, the eye of the sea-god decorating the root of the bow-ram, powered by two banks of oars and square sails of linen latticed with leather. Their crews were predominantly free men, tough oarsmen expected to fight like soldiers as well as be seafarers. Among the auxiliary vessels were longboats for inshore work. These were called *scaphae*, and a long line of descent connects the term directly to the scaffies predominant in the fishing in the early nineteenth century.

When the Romans first ventured into the North Sea and the Atlantic they were surprised by the tides and currents, a startling contrast to the almost tideless Mediterranean. Julius Caesar wrote of this in his account of the wars against the Gauls. Later the navy lost many ships in a storm on the German coast. By the time of Agricola, they must have been wary not only of encountering hostile natives but of the natural hazards they

Sandbars at the mouth of the Spey.

faced. They may also have learned from local designs for seagoing vessels, creating the beginnings of a Romano-Celtic boatbuilding tradition that may have become familiar to the Picts. In the Firth, once they had rounded Kinnaird Head and come west, the Roman navigators may have found a scarcity of landing places. The contemporary *North Coast of Scotland Pilot* advises mariners against approaching several stretches of the Moray Firth coast 'without local knowledge'. The brooding mass of Troup Head with its attendant flocks of seabirds, rearing more than 350 feet before them, would have been enough to make the Roman seamen wary of shoals and reefs. Even on the ostensibly gentler sandy coasts of Moray, Nairn and Inverness lurk sandbars. We can be relatively sure of such generalities but we cannot be certain of the detail of the coast when the Romans arrived, as the level of the sea in the Moray Firth has been falling relative to the land for the last 8,000 years. The position of the maximum sea level is plainly marked around the inner Firth coasts by the ancient raised beaches. By AD 84, the year of Agricola's voyage, the sea level was about 2 metres higher than at present. The cliff walls in Caithness and Buchan are probably much as they were when the Romans sailed past but it is a very different matter in the inner Firth where the extensive beds of gravel and sand are in constant movement. In his book *The Evolution of Scotland's Scenery*, J.B. Sissons said the coastal storm ridges on the south side of the Firth are among the most rapidly changing landforms in the country. We will never know what the Romans saw but the evidence from more recent coastal change hints at the extent of the difference. The Culbin bar was migrating westward at about 14 metres a year in the 1980s.[2] The Findhorn river used to enter the sea about a mile south-west of the present village where part of the old course was occupied by the Buckie Loch, and the present river mouth was carved out in 1702 when the sand-spit was breached. The mouths of the Lossie and the Spey have also shifted over the years. The sweep of sand ridges east of Tain, the Morrich More, has grown north-eastward by possibly 2 miles since the time of Agricola's fleet, and Whiteness Head probably did not exist when the Romans arrived. Burghead was probably an island. As well as these continual shiftings caused by current and tide, storms can bring more dramatic change. A great tempest in the North Sea between 120 and 114 BC flooded the coasts of Jutland and north-west Germany, and there was a similar although lesser incursion in AD 500. We can surmise that our side of the North Sea has also been shaped in such a way.

Around 50 years after the voyage of Agricola's fleet, in the city of Alexandria a scholar of geography called Claudius Ptolemaeus, better known to us as Ptolemy, compiled a description of the known world. It includes the earliest place names we know for the Moray Firth. Although he would have had access to information from other writers and travellers, probably to as far back as Pytheas, the Phoenician from Massalia, now Marseilles, who had sailed around Britain some four centuries before, the geographer's sources may well have included Agricola's seamen.[3] At the extreme end of Scotland, Ptolemy named three headlands – Tarvedrum or Orkas, Virvedrum and Verubium. The most likely identification for Tarvedrum, with its alternative name of Orkas, for Orkney, is Dunnet Head. Virvedrum and Verubium are probably Duncansby Head and Noss Head. These three from the point of view of a mariner passing along the coast are the significant waymarks.

The curling spit at the entrance to Findhorn Bay clearly shows the patterns of movement of the sand.

The next feature Ptolemy records in a north to south sequence is the river Ila, without doubt the Helmsdale or, in Gaelic, the Ilidh. South of the Ila lies what Ptolemy calls Alta Ripa, literally 'high bank'. Some have taken this to mean the Ord of Caithness and others have argued it is Strath Oykel but it is clear that, if once again the coast as seen by a mariner is imagined, the high bank is most likely the straight stretch of coast from the Sutors of Cromarty to North Kessock. This would fit with the likelihood that southbound sailors with a fair wind would strike for Tarbat Ness from south of Helmsdale to round the peninsulas of Easter Ross and the Black Isle before sailing up the Firth to reach the next place Ptolemy names, the estuary of the Varar river, to us the Beauly Firth, the gateway to Strathfarrar. The southbound tidal stream in the Moray Firth runs at a right angle away from the coast at Brora and rock ledges extend offshore

for some distance from Brora Point, two good reasons for a fleet to bear away from the coast here, and entirely miss Loch Fleet and the Dornoch Firth, two very obvious features now but which may have been less so two millennia ago. Ptolemy omits to mention them. To the east of the Varar estuary, Ptolemy named the Loxa river, possibly the Nairn, the Findhorn or the Lossie, where there would have been a much larger body of water in Roman times if Burghead was indeed on an island and the sea extended across the low land that later surrounded Loch Spynie. There is general agreement that in Ptolemy's nomenclature the estuary Tuesis must be the Spey, and the river Caelis is probably the Deveron. It is curious that the Sutors, the high guardians of the mouth of the Cromarty Firth, have been overlooked, unless they are disguised within the description of a 'high bank'. Perhaps the Romans had good reason to sail by them and therefore fail to include them later in relaying information to geographers. The spacing of the names hints that they may have been waymarks or anchorages. The mouth of the Helmsdale River, for example, would have been a welcome

The straight, high coast of the Black Isle and Easter Ross, cut by the entrance to the Cromarty Firth, fits the description of 'high bank', *alta ripa*, used by Ptolemy.

11

The great sandbars off the coast at Culbin. The coniferous forest dates from the planting by the Forestry Commission from the 1920s that successfully stabilised the notoriously shifting dunes of the area. The estate of Culbin was buried by blown sand between 1693 and 1695, forcing out the occupants, an extreme case of a process that continually threatened fields and farmyards alike in this district until recent times.

sight after skirting the long rock wall of Caithness. What has come down to us are from a mariners' viewpoint some highlights of the topography, the prominent features. Possibly some were also frontiers, between one tribe and its neighbour, signposts to the traveller to be prepared for difference, or they were markers for major connecting routes.

There has always been argument over how far north contingents of the Roman army advanced on land, argument kept alive by continuing archaeological research and reassessment of the evidence. The most northerly Roman marching camps so far identified with certainty lie in Glenmailen (Ythan Wells) in Banffshire and at Auchinhove near Huntly. Other camps – at Durno on the bank of the Urie and at Kintore – mark a line of advance skirting the edge of the central mountains. Smaller scouting parties of Roman infantry may well have come farther into the unknown. There is a possible camp at the mouth of the Spey, and other candidates for camp sites have been found at several places in the inner Moray Firth, for example at Delnies, near Galcantray, at Blar na Coille near the mouth of the Beauly river, and on the Tarbat peninsula.[4] Ptolemy refers to a Roman base called *castra alata*, an obscure name meaning winged camp, that some believe to have been Burghead.

The Alexandrian geographer peopled this far-flung corner of the world he would never see. To the east of the Caelis river, he identified the promontory of the Taezali. Who were these residents in what is now Buchan? And who were the other tribes he placed on his map – the Caledones, the Vacomagi, the Decantae, the Smertae, the Lugi and the Cornovii? The paucity of information in the old sources leaves room for endless argument about the actual territories these tribes occupied and the possible meaning and significance of the names. What can we make of the three 'towns' that Ptolemy assigns to the territory of the Vacomagi? One is called Tuesis and could well have been a centre of population or local power in the valley of the Spey, sharing for the Romans the name of the river. The other two, Tameia and Bannata, appear to have been further south and possibly even in Strathmore.

A quiet afternoon on the Beauly Firth, Ptolemy's Varar estuary.

What do we know of the environment around the Moray Firth 2,000 years ago? Although the climate at the time was slightly milder and drier than it is now, the natural vegetation would have been familiar. On the coasts under the scour of the salty winds, grassland would have been predominant, with poor scrubby woodland in more sheltered spots, giving way to denser

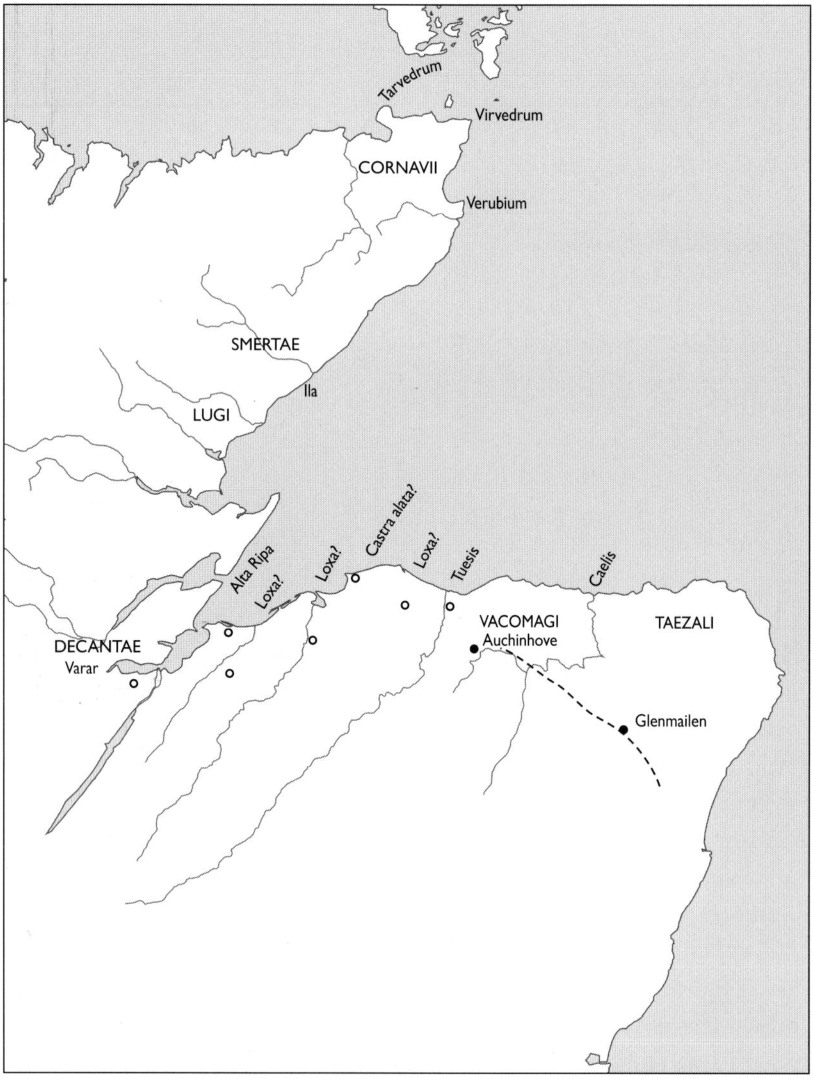

Suggested locations
for the places and
tribes named by
Ptolemy. The possible
line of advance of
Roman forces on land
through Glenmailen
to Auchinhove is
also shown. Other
candidates for
Roman camps on
the south side of the
Firth are indicated by
the open circles.

Suggested locations for the places and tribes named by Ptolemy. The possible line of advance of Roman forces on land through Glenmailen to Auchinhove is also shown. Other candidates for Roman camps on the south side of the Firth are indicated by the open circles.

forest of pine, hazel, birch, alder and oak according to the local conditions imposed by drainage, soil, altitude and exposure. This natural ecology had already been affected by centuries of human settlement before the arrival of the Roman fleet, but the impact must have varied greatly from place to place. Most of the people lived, as one might expect, in a coastal zone of varying width and the straths extending inland along the more hospitable stretches of the main river valleys where an arable economy was possible. Much of the expanse of the high country from the eastern foothills of the Cairngorms to far beyond the Great Glen may have had no permanent human population of any size. The distribution of archaeological sites in the Cairngorms that date from before AD 1000 shows permanent settlement limited to the major valleys, although flint arrowheads and other finds confirm that hunters exploited the upland country as well, as we might expect.

Can we reach any conclusions about the size of the population? Seventeen centuries were to pass after the time of Agricola before anyone made a systematic attempt to enumerate Scotland's inhabitants. This was done by the Revd Dr Alexander Webster in 1755; he carried out his survey by writing to all his fellow clergy to ask them for the numbers of men, women and children in their parishes and concluded that the population of the country was 1,265,380, a conclusion broadly supported by a survey repeated in a similar fashion for Sir John Sinclair's *Statistical Account of Scotland* some 50 years later. Webster's census, done before industrialisation and agricultural change had set in motion much migration from the countryside, shows a more equable distribution of people than is the case in Scotland now. About 27 per cent of Scotland's people lived in the counties bordering the Moray Firth and it is possible that this situation had pertained for a long time. We can take a risk here and propose it was so even in Roman times: that roughly a quarter of the population within the bounds of what would one day be the nation of Scotland was to be found around the Moray Firth.

The population of Scotland has been estimated to have been about 250,000 in the late eleventh century but it could have been twice that figure – we have no way of knowing.[5] The compilation of the *Domesday Book* after the Norman conquest allows an estimate of around 2 million for the population of England in the same period, but we have no comparable source of information for Scotland. Applying Webster's distribution pattern

This enigmatic, solitary stone at Ospisdale on the Dornoch Firth is a reminder of a long, mysterious past.

15

to a 250,000–500,000 estimate of total numbers gives a circum-Moray Firth population of 62,500–125,000. What was it, when the Romans arrived, a millennium before that period? We can assume no steady pattern of growth in population; we have to allow for major setbacks caused by warfare, natural disasters and epidemics and for longer-term fluctuations driven by slow-acting influences. For example, we have the evidence for a widespread climatic shift to mild, wet conditions and a subsequent spread of blanket bog around 2000 BC that drove people from higher ground to escape a marginal existence and the risk of famine.[6] Peat bogs and tree rings also bear the marks of large volcanic eruptions, for example of Hekla in Iceland, affecting weather and vegetation in Scotland, events that would have mystified Neolithic people as much as they imperilled their agriculture. The Roman world was swept by an epidemic of plague between AD 165 and 180, called the Antonine Plague after one of its most prominent victims; a burial pit associated with it has been found in Gloucester but we have no evidence that it reached Scotland, although of course this does not rule out the occurrence of other unrecorded outbreaks of disease.

We are on slightly firmer ground in arguing that most of the people in the Firth area would have lived along the southern coast, where the better arable land is to be found. The modern *Soil Survey of Scotland* maps the best arable ground as being on the peninsulas of Easter Ross and the Black Isle, in patches along the valley of the Nairn and in broader swathes through the Laich o' Moray from Dyke east to Duffus. A larger acreage of slightly poorer ground but still good for cereals and grass spreads around the best spots in an extensive sweep from Buchan to Brora, and further to the north in a belt between Wick and Thurso. Much of the better land is underlain by Old Red Sandstone rock. These desirable acres are fringed everywhere by zones of more sour or rocky soil that gradually give way to moorland. Though farmers have through centuries of tillage changed the soil that nature has presented to them and have generally expanded the area of cultivation, the underlying distribution of natural fertility is probably well-enough reflected in our contemporary soil maps. What appeals to the farmer now is likely to have always had attraction.

The inhabitants of the Moray Firth area in Roman times were a people engaged in farming, with some fishing and hunting, scattered in small settlements and individual farmsteads probably based around kindred groups and forming small tribes. A settlement may not have been too unlike the more familiar Victorian fermtoun, a small community drawing

One of the best preserved brochs on the northern mainland is this one at Strathsteven, called Carn Liath, between Brora and Golspie. The visitors convey the dimensions of the central circular chamber, and the space between the double wall is also evident. A complete broch has a shape that has often been compared to an industrial cooling tower but none of the brochs around the Firth has survived to this extent.

A characteristic building for the far north around the time of the Roman voyage in AD 84 is the broch. Further south the vitrified fort is the predominant large structure. This distribution of brochs and forts has been interpreted to reflect ethnic difference between the regions, with a boundary coinciding on the east coast roughly with Strathoykel and the Dornoch Firth, but it could reflect more the availability of good building material influencing local customs. The view that brochs were built as defensive strongholds in a troubled time has also been challenged – they may have been erected as declarations of status, expressions of wealth in stone. Other building types have been commonly found, such as souterrains and, where there are bodies of water, crannogs.

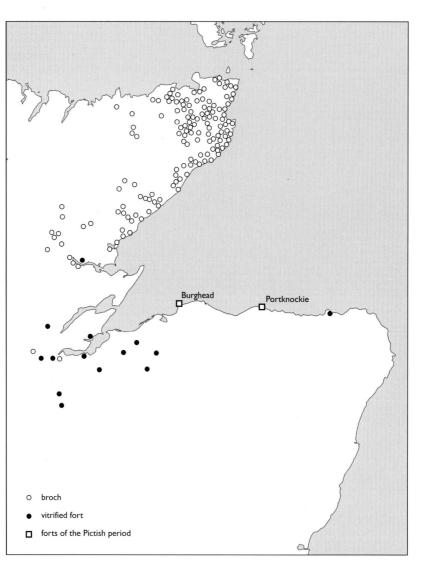

its living from a distinct area of country, knowing every inch of its ground and what it could offer. There would not of course have been the same obligations and social ties as in modern times, but there would have been equivalent social structures, equivalent imperatives, and in a few locations a number of farmsteads may have run close enough to each other to make up more populated, more complex settlements. The society around the Firth was certainly Celtic and no doubt had much in common with the Celtic culture that occupied most of central and northern Europe.

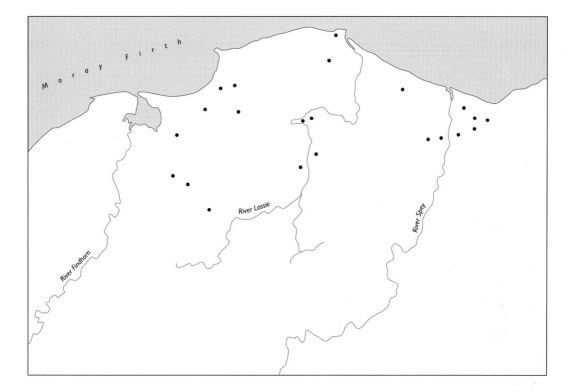

Tacitus wrote that the Celts under their leader Calgacus mustered 30,000 men to face Agricola's invading army at Mons Graupius, probably the outcome of the tribes uniting in some kind of federation against a common threat. We cannot say from what distances men gathered to fight under Calgacus but most would have likely come from the districts closer to the battlefield – south of the Cairngorms, if the conclusions of most scholars are correct. Tacitus records that one-third of their number fell. The effect of such a loss of active males must have been devastating and taken a long time to fade. Apt was the accusation Tacitus put into the defeated Caledonian leader's mouth: 'You Romans make a desert and you call it peace.'

After their voyages, their scouting trips and their short-lived military presence, the Romans abandoned any intention to expand the empire into northern Scotland. They probably judged the economic rewards to be not worth the effort, and there may have been a sense as well of their power reaching its natural and most easily sustained limits. The Emperor Hadrian set a policy of strengthening and fortifying existing frontiers after AD 117.

The distribution of brochs and forts could give a misleading impression that the lowlands of Moray were sparsely populated. The opposite is more likely to have been the case, as the map of crop marks, enclosure marks and other evidence for farming in the Iron Age shows for the low country between the Findhorn and the Spey, where presumably building was predominantly in wood. The recent discovery of Iron Age roundhouse settlements at Inverness, Birnie and elsewhere support this view.

19

On the Continent, the Rhine was recognised as the edge of the empire and, in Scotland, the Romans retired from probing into the Caledonian hills to a demarcation line running across the waist of the country along the southern slopes of the Kelvin and Carron valleys, a line that came to be known as the Antonine Wall. This construction, of turf and clay on a stone base, seems to have been actively occupied for only some 20 years, from about AD 142 to the 160s, before the Romans withdrew further south to the more strongly defended Hadrian's Wall. Thereafter, apart from a few campaigns, most notably one led by Septimius Severus in AD 208–10, the inhabitants of the north were left outside the empire. The abundance of Roman objects found in northern archaeology shows, however, that although there may have been around the Moray Firth few if any Roman feet on the ground, the empire was having a far-reaching influence on local life, rather as the impact of European imperialism in recent times reached across the world beyond their colonies.

# CHAPTER 2
# EMERGENCE OF A PICTISH NATION

The collapse of the Roman Empire created a political vacuum that was soon filled by a host of power centres across the Continent of Europe. Among them were the small kingdoms that made up what we know to be the territory of the Picts. The term *Picti* in reference to the inhabitants of Scotland north of the Forth–Clyde axis is not recorded anywhere until the year 297, when the words of a Roman orator and teacher called Eumenius about the 'Caledonians and other Picts' were noted down in what is now Autun in France. Thereafter the name occurs several times in Latin sources, most strikingly on a decorated dice tower from an excavation of a Roman rural settlement near Cologne. The inscription on the dice tower, probably of the fourth century, reads *PICTOS VICTOS HOSTIS DELETA LUDITE SECURI* – 'the Picts are beaten, the enemy annihilated, let us play without care', a slogan that tells us the Picts were famed for good or ill far beyond their native shores.[1]

This reconstructed Iron Age village at Butser in Hampshire is a long way from the Moray Firth but the settlements of roundhouses built from wattle and daub roofed with thatch may not have been dissimilar in appearance.

Left: The Invergowrie stone with its image of a drinking Pict. Right: The Portmahomack stone.

There is still some debate over what is meant by *Picti* – in Latin, literally 'the painted ones' – but it was probably a derogatory nickname the Romans derived from the habit the natives had of tattooing their bodies and used generally for their 'enemies' beyond their northern frontier to distinguish them from the civilised Britons within the Roman province of Britannia. Our name for ourselves – Scots – comes from another Roman slang term, the Scotti, their nickname for pirates raiding from across the Irish Sea. The Northumbrian cleric, the Venerable Bede, writing in 731, put down a story that the Picts had invaded the north from the land of Scythia but this explanation of their sudden appearance is a myth. In our time, the Picts have also been mythologised in the popular imagination, labelled a problem or an enigma, endowed with an air of mystery and placed at the focus of a vast literature that has deployed them to suit changing viewpoints. They are remembered where many other ethnic names used by the Romans have been forgotten – who, for example, considers the Attacotti enigmatic or problematic, and they are a Scottish tribe mentioned only once in Roman sources as allies of the Picts and Scots in the year AD 367. The hard evidence about the Picts, from surviving chronicles, archaeology, place-name studies, and other disciplines, points to a more common-sense explanation – that the people the Romans called Picts were the descendants of the people Agricola's army encountered in AD 84, tribes occupying Scotland from the central belt to Shetland, well established in their homeland, and in their way of life broadly similar to the other Celtic peoples the Romans knew across their empire. There has been much argument over the language or languages the Picts spoke, fuelled by the mysterious ogam inscriptions on many of their stone monuments, but it seems plain that they spoke a P-Celtic tongue, akin to the other Celtic languages spoken in Britain, now known as Brythnoic.

The Cadboll stone.

The Pictish economy was grounded in agriculture, an agriculture about which we have only a general knowledge. Archaeologists have pinned down and excavated very few farmsteads dating to the Pictish period, most likely because later farms have been established right on top of them, obscuring what may have survived from generations of occupation and tillage. Mapping the field systems that can be dated to AD 0–800 shows that, naturally enough, farming was concentrated more or less on the same stretches of land as it is now. One farmstead, at Buckquoy on the north side of the Bay of Birsay on the Orkney mainland, has been excavated and described in detail.[2] The most tangible legacy of the Pictish period is of course the array of stone monuments with their intriguing symbols. In Aberdeenshire, the older monuments, from the sixth century or earlier, tend to be on poorer soils whereas succeeding classes of monument from later centuries appear on better ground at lower altitude. This is taken to be some evidence for growth in population and an expansion of tillage to heavier soils on lower ground.[3] About half the animal bones found at Buckquoy are from cattle, of a breed about half the size of their modern descendants and akin to Dexter cattle. Sheep were also common, sheep like the mouflon at first and later like the Soay breed, and one-fifth of the bones are from pigs. The assemblage of bones includes in small quantities the remains of a wide range of other mammals – goat, horse, dog, cat, deer, seal, whale, otter, rabbit, mouse, vole and rat – together presenting us with a picture of a community actively engaged with livestock and hunting, as well as having to thole the presence of common inquilines and pests. The bird bones suggest that the Picts, at least in Orkney, had no domestic fowl but hunted and harvested seabirds, geese and waders. They also fished and gathered seafood from the shore. These are activities that have always been associated with life in the islands and it is no surprise to find traces of them among the Buckquoy sediment remains. At Buckquoy about one-third of the animals had been killed in their first year of life, slaughtered for food. Bones last better than vegetable remains in the old middens now turned to soil and give a false impression of a high protein diet, but grain was almost certainly the staple item on the Pictish table, with a variety of legumes and other plants, as witnessed by surviving carbonised seed. The most thorough exploration of Iron Age agriculture in Britain has been carried out at Butser in Hampshire, the opposite end of the country from the Moray Firth but, with some speculative allowance for the difference in latitude, soil and other influences, what has been discovered in the south probably applied to some extent in the north as well. Barley, wheat and oats would have been the main cereals and experiments have shown that the Iron Age ard ploughs, fashioned from wood, with an iron sock on the business end, and drawn by oxen or horses, were more efficient tillage implements than their humble appearance might suggest. In the south of England storage pits for grain were commonly dug in the chalk bedrock. Two grain storage pits have been found in only one location in the Firth area, in gravelly soil in the parish of Rosskeen on the north side of the Cromarty Firth,[4] and above-ground storage in barns was probably the northern norm. Studies of yields from Iron Age crops have shown that the arable farming of the time could provide a reasonable sufficiency, certainly in good years comfortably above subsistence level.

We can imagine around the Moray Firth a network of kindred groups and tribes, farmers and hunters, tied closely to the land and the seasons. Such a society would have had its hierarchies, its leaders and followers, and a system of law and custom to regulate its obligations and disputes. There would have been levels of status, probably with what could be termed 'free', powerful families at the top with command over subordinates with varying degrees of independence and, at the bottom of the social heap, a class of 'unfree' followers and slaves. The stone monuments present us with some characters and scenes from Pictish life. Some of the figures have a cartoon-like charm – the bearded rider draining his upended drinking horn with its end carved to suggest an amused bird while his horse plods on head down, on the Invergowrie stone – while others record more serious pursuits such as hunting and fighting. There are plenty of long-haired warriors, cloak- or tunic-swathed, mounted or on foot, with square shields, spears, bows or swords. Such may have been the society, evolving at its own pace according to its own rhythms and needs, that found itself having to cope with the Roman imperium. The effect of military incursion and defeat was one obvious and dramatic impact but, as far as we can judge from the surviving records, these reverses were few and far between for many years after Mons Graupius. Trade with Rome also played its part and in time drove more significant, lasting change. Exactly how this change happened will probably always remain unknown but some reasonable speculation is possible. Across northern Europe the leading kindreds, the social elites, among the so-called barbarian tribes were affected in much the same way. Trade with Rome brought new kinds of commodities, goods that appealed to the elites, who had the means to acquire them and probably the first pick of what came their way.

In 1996 a metal detectorist exploring a field at Birnie near Elgin came upon some Roman silver *denarii*, tiny coins stamped with the images of authority. Further investigation and excavation by archaeologists from the National Museums of Scotland uncovered, among other things, a hoard of coins, more than 300 of them in the remains of a leather bag in a broken pot that had been lined with bracken. The latest coins in this mysterious accumulation of Roman money date from around AD 194. The pot was locally made and it was dug from a pit close to a large native roundhouse, one of a number of dwellings in a small village. Not long afterwards, another hoard was found, buried in a pot in a similar way about 10 metres from the first one. There was no cash economy in the north at the time and much thought has been directed at the value the owner or owners of the hoards could have placed on the coins. The general conclusion is that Birnie was the home of a local power broker and the coins were in his possession as a bribe, a subsidy or some symbol of status.[5]

Such a status symbol might easily have attracted envious attention. It is thought that this unequal acquisition of status and 'luxuries' had the consequence of destabilising the social structure, probably straining obligations and enhancing rivalries. An outbreak of plague or a bad harvest would have made any destabilising effect even worse. Leaders with access to the new goods could also attract and retain adherents, followers who became in time bands of professional fighters, living at the leader's expense, shunning the daily grind of agriculture and forcing the leaders in turn to find ways to keep them under control, in effect to keep them occupied, fed and content. There probably resulted increased warfare between tribes and a tendency for some kindreds to ascend in power and wealth at the expense of their neighbours. The development of this phenomenon may have been one factor behind the raids that the Picts began to inflict on Roman Britain. Raiding appears to have grown more common towards the end of the fourth century: for example, a *barbarica conspiratio*, as the Romans have it, from the north ravaged the province of Britannia in 367, and 17 years later the Romans were again fighting the northern tribes, this time a joint incursion by Picti and Scotti. These incursions were probably a response to the weakening of Roman power but they may also have grown from the growth of military capacity among the tribes and a keenness to find new sources of plunder to support the prestige of the leaders.

Over time many of the smaller, weaker kindreds became absorbed into larger political units, and leaders began to be called 'kings' in the contemporary chronicles and annals as a more universal Pictish identity supplemented local loyalties. Some manuscripts from the Middle Ages record how the land of the Picts was divided. The topography imposed some obvious boundaries. The Cairngorms, the Mounth in the old chronicles, separated northern Pictland from its southern counterpart but smaller divisions were also recognised. In *De Situ Albanie*, a manuscript dating from the mid twelfth century, seven brothers, the sons of Cruithne, whose name means simply Pict, are said to have shared Pictland between them and have given their names to the regions they ruled. There are several versions of this tale – some name the brothers, one set of these founding siblings being Cait, Ce, Cirig, Fib, Fidach, Fotla and Fortrend.[6] It is legend but there is reluctance to accept that it has no basis at all in geographical fact, with Cait obviously Caithness and Sutherland, Fib Fife, Fotla Atholl and so on. In this scheme of things, Fortriu – Fortrend is the genitive case, meaning 'of Fortriu' – was identified more or less by elimination by A.O. Anderson as the Strathearn/Menteith region and Fidach as Moray. More recently, in a general reassessment of the Dark Ages in Scotland, a good case has been made for Fortriu being Moray.[7] This province was much more extensive in the Middle Ages than the present county with the same name. According to Lachlan Shaw, it 'extended from the mouth of the River Spey on the east to the River Beauly on the west. A line . . . from Loch Lochy . . . through Lochaber and following nearly the course of the River Spey along the base of Cairngorm and Ben Rinnes formed its southern boundary.'[8] The conclusion that this great territory was the heart of the old Pictish kingdom of Fortriu rests upon a detailed analysis of the source material, a hallmark of scholarship about the Picts but beyond the scope of this volume.

The emergence of the Pictish kingdoms was part of a general pattern found across much of Western Europe as new political orders struggled from the ashes of the Roman Empire in long periods of dynastic conflict, only glimpsed in the terse lines written down mainly by clerics in their scriptoria. The Christian church burgeoning from its beginnings as a persecuted sect within the Roman Empire to become the first pan-European institution was intimately wrapped up in these political processes as an ally or otherwise of secular rulers. The church could bestow or withhold its blessing; an ambitious dynast could see the importance of having the church and God on his side, and so rulers could either advance the church through grants of land and affording it protection or refuse it house room.

The arrival of Colum Cille, or Saint Columba as he is usually known, on Iona in 563 is often celebrated, and rightly so. Christianity was already being practised in the south of Scotland but, when Colum Cille and his 12 companions stepped up the beach of the small island that would for ever-more be seen as the cradle of the new religion and a place of tremendous spiritual importance, the Picts around the Moray Firth still adhered to their old beliefs. Colum Cille was born in Donegal into one of the ruling families in the Gaelic world that straddled the North Channel. He was destined from the start for a career in the church – he was intimately involved in the founding of monasteries at Derry, Durrow and possibly Kells – but he was never fully separated from more earthly power struggles and was linked in some major, though now obscure, way to a battle between rival

Irish dynasties. A realisation that it might be wise to be out of the country for a while may well have been part of the motivation for the mission to the north in 563. The entries about Colum Cille in the various annals are typically matter of fact and unrevealing – 'The sailing of Colum Cille to the island of Iona in the forty-fifth year of his age', is all the Irish chronicle, the Annals of Tigernach, has to say.[9] Much of what detail we have about the man comes from the account of his life written by Adamnan, one of his successors as abbot of Iona. This is more hagiography than biography, as Adamnan is interested only in telling how his subject was a great Christian. It has memories of the blessed man and anecdotes of miracles performed, healings made, conversations held, all in various parts of the country and following no chronological sequence. According to Adamnan, a few years after establishing himself in Iona, Colum Cille ventured across the 'spine' of the country (dorsum Britanniae) to his famous encounter with Brude, son of Maelchon, the king of the northern Picts. Only once does Adamnan refer specifically to Colum Cille's first journey to Brude. It reads as follows, in the translation by William Reeve:

> . . . when the saint made his first journey to King Brude, it happened that the king, elated by the pride of royalty, acted haughtily and would not open his gates on the first arrival of the blessed man. When the man of God observed this, he approached the folding doors with his companions and, having first formed upon them the sign of the cross of our Lord, he then knocked at and laid his hand upon the gate, which instantly flew open of its own accord . . . and when the king learned what had occurred, he and his councillors were filled with alarm . . . and he [Brude] advanced to meet with due respect the blessed man . . . and ever after from that day, so long as he lived, the king held this holy and reverend man in very great honour . . .[10]

Adamnan makes it clear that Colum Cille made several trips into the province of the Picts and mentions twice the Ness, whether river or loch, but otherwise does not tell us where the Pictish power centre lay. Recent re-evaluation of Pictish history has brought forward the suggestion that the great event may even have occurred in Strathtay.[11] In this scenario the Pictish realm of Fortriu is assumed to be 'diphyletic', split over two regions, the Firthlands around the Moray Firth and Atholl on the other side of the Cairngorms. Again according to Adamnan, Brude embraced Christianity after Colum Cille's humiliation of Broichan, the arch priest of the pagan religion of the Picts. All Brude's subjects almost certainly did not follow suit, at least not immediately, and even Brude himself may have kept private pagan counsel while outwardly adopting Christianity for some political advantage. The ability of a people to conform to new practices, especially those foisted on them by their superiors, but not abandon the old is amply confirmed by the reports from a later religious revolution, the one we call the Reformation.

A considerable missionary effort sprang from Iona and penetrated to all the corners of the north, continuing and flourishing long after Colum Cille's death in 597. Many of these Celtic clerics are remembered as saints, for example, Donnan, who founded a monastic centre on Eigg and met a violent death there at the hands of robbers; and Maelrubha who established his monastery and sanctuary at Applecross, but who also was adopted as the patron of Keith, originally called Cet Maolrubha – Maolrubha's wood.[12] Linked to the Moray Firth area rather than the west are Fergus, active in the north-east and in Caithness in the early eighth century, and Comgall, who preached to the Picts along with Colum Cille and died in 601. Such was the power of Colum Cille's name that he was invoked as the founder of the monastery of Deer, where Drostan was the first abbot and died in the early seventh century.[13] A female saint, Bey, who was active around 816, is associated with Banff. A prominent later saint was Duthac: he died and was buried in Tain in 1065 but when his body was discovered to be uncorrupted by decay it was installed in a shrine that thereafter became a major centre of pilgrimage in Easter Ross. Some of the monastic sites linked with these early saints and preachers perhaps lie silently under later kirks or graveyards that preserve the spirituality associated with their founders. This could be true of, among others, Kinneddar, Birnie, Spynie, Mortlach and the Old High Church in Inverness where tradition has it Colum Cille himself preached to the assembled subjects of Brude.

The discovery in the garden wall of the Tarbat manse at Portmahomack of a stone slab carved with a Latin inscription led to the recent, very important excavations of what was a Pictish monastic centre in that part of Easter Ross. The dig has uncovered stirring detail of life in a Columban religious foundation, with relics of vellum making, metalworking, sculpture and agriculture. This centre was destroyed by fire sometime in the eighth or ninth centuries, but later in the twelfth century became the site of the

Tarbat Old Parish Church at Portmahomack dates back to the twelfth century but overlooks the site of a monastery founded in the late sixth – early seventh centuries during the Pictish period. This was a centre for the making of books and artefacts, metalworking and stone carving, and the archaeological remains include the foundations of a large eighth-century barn. The monastery was destroyed by fire, possibly the victim of a Norse raid.

parish church of Saint Colman. The monastery lying beside the strategic ship portage across the neck of the East Ross peninsula may well have been the place where craftsmen carved the cross slabs at nearby Nigg, Shandwick and Hilton of Cadboll. These monuments combine Christian imagery with natural or abstract symbols. The most impressive tangible legacy the Picts have bequeathed to us is their stone sculpture. The symbol stones continue to intrigue us with their expertly depicted animals – boar, salmon, eagle – and their mysterious combinations of abstract images. There is in them an affinity to Celtic, Saxon and Norse art but they continue to defy interpretation, despite the many ideas that have been put forward to explain them. Several attempts have been made to classify the Pictish stones but there is consensus that they show a progression in technique and in inspiration from pagan to Christian.[14] Very few of them can be dated very precisely: they may first have appeared in the fourth century, and some scholars have argued for them continuing to be made as late as the tenth or even the twelfth century, long after the Picts had been absorbed into a Gaelic hegemony. The distribution of the sites where the symbol stones have

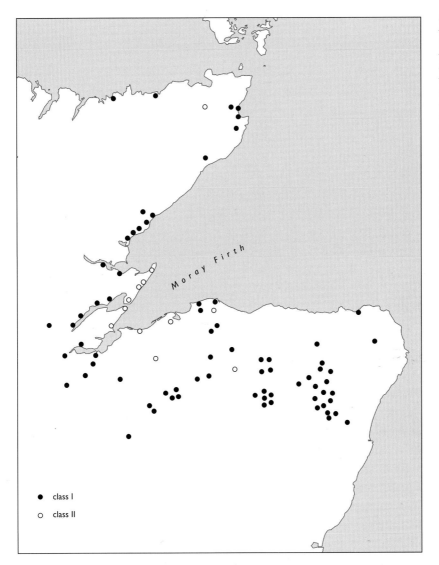

On this map showing the distribution of Pictish stones around the Firth, the solid circles indicate those with symbols only and the open circles those with symbols and Christian crosses, Classes I and II according to Carver's classification (Carver, 1999). Most of the Christian stones cluster within reasonable distance of the Portmahomack site.

been found suggests that this distinctive Pictish art may have originated around the Moray Firth, and it has also been suggested that somehow the Romans influenced their origins. This Moray Firth origin may have been one aspect of the flourishing of Fortriu as a regional power. The earliest construction at the great fortress of Burghead has been dated to the third century, around 260, with a similar date for the stronghold at Cullykhan, near Pennan.[15] These dates fit with the proposal that a centralising tendency, a coming together in the face of an external threat, came about

Image of a Pictish ship, from Jonathan's Cave, East Wemyss, Fife.

under the impact of the Romans, a tendency that would have stimulated the emergence of Fortriu and its neighbouring kingdoms. The power of Fortriu reached to include Caithness and Sutherland, easily visible from the Burghead coast, and Strathspey and probably a considerable distance down the Great Glen. Although the Cairngorms may have shielded the north from some invasions, the Pictish kings in our region were nothing loath to strike at their neighbours, as is made starkly clear in the monks' annals. According to the Annals of Ulster, one of the powerful kings of Fortriu, Aedan, ruled from 576 until about 609. In or shortly after 580, he launched a campaign in Orkney. From records of this sort we deduce that the Picts knew the sea and had a navy, although there remains much unknown about their ships. Clearly this navy could cross the turbulence of the Pentland Firth and deliver an armed host to a beachhead. Burghead, Cullykhan and Green Castle at Portknockie may represent bases separated by convenient sailing distances. The Annals of Tigernach record that in 729 no fewer than 150 ships were wrecked in a storm at a place called Ros-Cuissine, which has been taken to be Troup Head. Such a casualty figure is reminiscent of the losses inflicted on fishing fleets in the Moray Firth in the nineteenth century and suggests that most of the Pictish 'ships' may have been of a comparable size. In the south Aedan's power extended to

Argyll, parts of Antrim and lands around the Forth before his armies ran up against those pushing north from Bernicia, the Anglo-Saxon kingdom that extended from the Humber, whose king, Aeðilfrith, defeated Aedan at the battle of Degsastan in 603.

Bridei, who died in 693, is another Pictish king to stand out from the annals as a significant figure. Indeed he is the first in the Irish annals to be called *rex Fortrenn*, king of Fortriu. His father had been ruler of the British kingdom centred on the Clyde, his mother was of Deiran descent, a Saxon from south of the Tyne, and he was also a cousin of the Anglo-Saxon Ecgfrith of Northumbria, such were the relationships forming as dynasties sought advancement through marriage as well as on the battlefield. Northumbria, formed from the amalgamation of the two Anglo-Saxon kingdoms of Bernicia and Deira, continued to push northward into Pictish territory in the late 600s, and Bridei probably had his first

The high bulk of Troup Head dominates the south-east corner of the Firth. This is possibly where a fleet of Pictish boats was wrecked in AD 729.

33

experiences as a warrior grappling with the Anglo-Saxons in the woods and mires of the Forth valley. In 680 he laid siege to Dunottar; in 681 he attacked Orkney, possibly to put down an insurgency against his regime; and in 683–84 he campaigned into the south and besieged the fortress of Dundurn in Strathearn. His greatest victory, and the one with most significance for the future of Scotland, came in 685. Usually known as the battle of Nechtansmere, this conflict stopped the northward expansion of Northumbrian power under Ecgfrith. The site is named as Dun Nechtain in the Irish annals and has for long been associated with Dunnichen near Forfar but recently a strong case has been made for the battle having been fought at Dunachton in Badenoch.[16] Bridei died in 692 and was buried on Iona in the grounds of the mother church that had flourished along with his realm. His victory at Dun Nechtain heralded the apogee of power for the Pictish kingdom of Fortriu. Although it now extended from Orkney to the valley of the Forth, its heartland came to be in the Tay valley, a shift in the political centre that was to have consequences for the Moray Firth area.

# CHAPTER 3
# INVADERS

What proved to be the most fateful threat to the Pictish realm around the Moray Firth came from the south-west, from the kingdom of Dal Riata. This polity extended from Kintyre north as far as the Ardnamurchan peninsula and the headwaters of Loch Linnhe, and included the islands from Mull round to the Firth of Clyde. The long-standing tradition is that this part of the Highlands was settled by Gaelic-speaking immigrants from Ireland in the fifth century, immigrants known by a nickname bestowed by the Romans on Irish raiders, the 'Scotti'. This word is of course the origin of the name of Scotland but, to avoid confusion with the later speakers of the Scots tongue, here we shall refer to the people of Dal Riata as Gaels.

Tradition credits the leader of the founding settlement from Ulster to have been Fergus, son of Erc, who died in 501, but the Gaels had been long in Argyll as the northern part of a realm spanning the North Channel. They designated themselves as the kindreds of their eponymous founders, such as the Cenel Loairn in Lorn and the Cenel nOengusa in Islay, where *cenel* means seed or descendants. Dunadd, the famous fortress built on the prominent rock outcrop in the valley of the Add river, in the approximate centre of the Dal Riata territory and dominating the obvious route across the neck of Argyll, served as a capital and ceremonial site. A collection of Irish manuscripts from the fourteenth century, the Book of Ballymote, gives a glimpse of this society in the south-west Highlands. The writer describes villages or districts with from 30 to 120 houses, and fighting strength of tribes as being only a few hundred men. The Cenel nOengusa could put 500 men into the field, the Cenel Gabrain 300 and the Cenel Loairn 700.

Although they spoke different Celtic languages, the general ways of life of the Picts and the Gaels were probably very similar – a core economy of agriculture practised in scattered settlements and farmsteads supporting a prominent warrior class. In 367–69 and again in 382 the two peoples formed an alliance to attack the Romans, serious incursions that at least in the latter instance took the Romans some time to overcome. Raids such as these – and there were more in the years following the withdrawal of occupying Roman forces – led Gildas, the British cleric on whose writings we rely heavily for an account of this time, to refer in about 540 to the northern Celts as 'foul hordes' crawling from the rocks like worms in warm weather.[1] The missionary work of Colum Cille and his followers also served to bring together the Picts and Gaels from the mid sixth century onwards. All of this, however, was not sufficient to prevent conflict eventually arising between the peoples, adding the dimension of ethnic struggle to the already violent dynamics of the interplay between rival dynasties.

The Anglican and the Celtic churches also became locked in a contest over whose interpretations of doctrine and practices would have official sanction in the British Isles. Celtic missionaries ventured from the west to bring their version of the Christian message to the Continent. Columbanus, a follower of Colum Cille, travelled in Gaul and crossed the Alps to die in a monastery he founded in northern Italy. The alternative power in the Christian world, under the leadership of the Pope, mounted a missionary effort in the opposite direction through the arrival in Kent of Augustine in 597. There were many points of difference between the rival churches but the one that is best remembered now is the dispute over the method for dating Easter, the central festival in the Christian calendar. Finally, in 664, in a synod at Whitby, the king of Anglo-Saxon Bernicia settled on the Roman dating. Other rulers followed the Bernician example, putting the Celtic church on the back foot. In 717 Naiton or Nechtan, the king of the Picts, expelled the monks and clerics of the Iona church from his territory after the followers and successors of Colum Cille rejected the king's ideas for reform, including over the dating of Easter. Some Celtic clerics continued to observe their own brand of the faith as hermits or as small groups of devotees; they called themselves Ceile De, servants of God, a term later anglicised as Culdees, and carried on serving their immediate communities.

Fortriu continued to be a strong, aggressive realm. In 728 victorious campaigns against the Anglo-Saxons in Lothian and the Britons in Strathclyde enlarged the Pictish territories. A setback was suffered after a defeat by the Gaels in 731 in a battle at an unknown location named as Muirbolg in the Annals of Tigernach – candidates for the location include Kentra Bay in Ardnamurchan, Loch Earn and the west side of

Loch Lomond, all on the fringes of Dal Riata[2] – but it was only temporary. The Picts returned to the offensive and by the early 740s the Pictish king, Onuist (Angus in Gaelic), held sway over the lands of the Gaels, at the head of a single, strong kingdom north of the Forth. Despite hints in the chronicles of internal dissension – for example, in 739 Onuist drowned the son of a king of Atholl, perhaps to remove a threat of usurpation – a relative stability persisted for a few decades until, after the death of Onuist in 761, the Gaels began to recover some of their former power.[3] The chronicles record a very severe winter in 763–64, followed by a famine and, in Ireland, an epidemic, and these natural disasters would also have sapped the strength of the combating Celts. Throughout this period, as at other times, natural occurrences – environmental setbacks of various sorts – probably had a strong influence on political affairs but it is seldom that we know much of what took place. Around 537 the *Annales Cambriae* note a plague in Britain and Ireland, and the Annals of Ulster record a 'great snow' in 589. In 700 there was a 'great frost', enough to freeze rivers and lakes in Ireland and, scarcely credibly, the sea between Scotland and Ireland so that communication took place over ice.

In 789 another strong, able ruler emerged victorious from a power struggle within the Pictish realm. This was Constantin, son of Wrguist, his name another manifestation of Continental Christian influence on Pictish society. Fate dealt him a testing hand, as he came to power in time to face a new threat from the north. For the year 794 the Annals of Ulster note 'The devastation of all the islands of Britain by the gentiles'. From such comments we have the classic image of the Vikings in their dragonships sailing out of the northern mists with a ferocity that terrifies the targets of their aggression. Within a handful of years, after raids on various islands and settlements, the Norse dominated the Irish Sea and probably much of the Atlantic seaboard. In 802 the monastic community on Iona fell victim to their depredations and four years later the Annals of Ulster lamented the murder of 68 monks of Iona in another raid.[4]

The Norse had superb ships and had learned to handle them on ocean-going voyages, a grasp of technology that gave the raiders the tactical edge. It is important to realise, though, that Norse ships were not entirely new but were rather a sophisticated advance on the type of ship found around the North Sea. Boat finds from various parts of northern Europe, from Ireland to the Baltic, show similarities in construction, but it appears that the Scandinavian shipwrights discovered the hull form that maximised speed. There must have been many accidents and drownings before the Norse brought their maritime skills to the level that allowed them to regard the north Atlantic as a sea road. We have seen already how the Picts were adept at sailing in their home waters, and it is intriguing to ponder on what course European history might have taken if it had been they who had resorted to blue-water voyaging and not their Scandinavian neighbours. That travel by sea in the vessels of the time was slow and dangerous did not prevent a steady level of contact with the Continent. The Moray Firth area was far from being an isolated remote corner of Europe. Shards of seventh- and eighth-century pottery from France have been found in many archaeological sites around the Irish Sea and north as far as Skye and Inverness. Monks of the Celtic church made the perilous crossing of the North Sea to bring their brand of Christianity to the Rhineland and further into the heart of the Continent. Known as *peregrini*, they travelled extensively along the existing trade routes and have left to mark their way a large number of monastic centres through western and central Europe. What was happening in the country of the Franks, where Charlemagne ruled over an expanding realm and reshaped the idea of kingship, was of intense interest to the minor kings in the Atlantic islands. In the life of Charlemagne written by the courtier and administrator Einhard, it is stated that 'the kings of the Scots [Irish] [were] so inclined to his will by his munificence that they never called him any name but lord, nor themselves but his subjects and servants' – a conceit on the part of Charlemagne's loyal supporters but an indication of the circulation of ideas as well as goods.[5]

The first parts of the British Isles to receive ship-borne Norse were Shetland and Orkney. This migration may have been continuing at a low level for a long time before the major assaults recorded with such horror in the Christian annals. What led the pagan Norse seafarers to spill from their homelands with such force has long been debated. There is unlikely to be a single cause behind the raids but dismal prospects at home must have been part of the story. The climate in north-western Europe seems to have grown colder and wetter in the few centuries before 800 and this would have led to poorer harvests.[6] A traditional reason is that a king called Harald the Fairhaired fought his way to supreme power in Norway, precipitating the emigration of those who could not thole his authority. 'In the discontent when King Harald seized on the lands of Norway, the out-countries of Iceland and the Faroe Isles were discovered and peopled. The Northmen had also a great resort to Shetland and many men left Norway, flying the

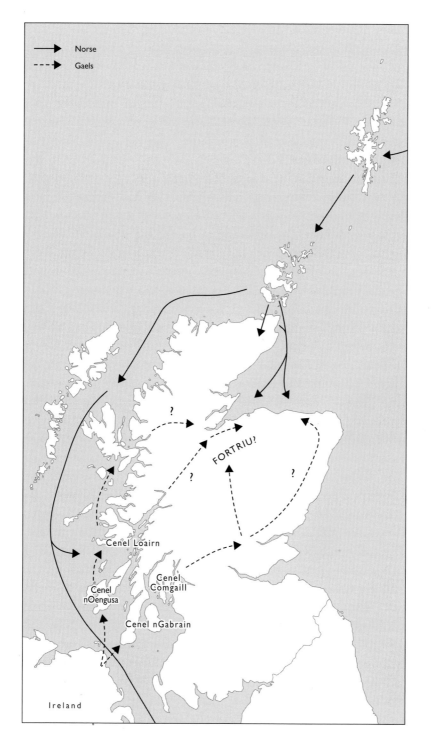

How the Pictish realm in the Moray Firth region was assailed from both north and south – from the north by Norse raiders and settlers, and from the south and west by Gaels moving out from their heartland in Dal Riata. The Gaels organised themselves into kindreds – the Cenel Loairn may have been the most important group to move into Moray, postulated to have been the centre of the Pictish kingdom of Fortriu.

Norse

Gaels

?

FORTRIU?

?

?

Cenel Loairn

Cenel Comgaill

Cenel nOengusa

Cenel nGabrain

Ireland

country on account of King Harald, and went on viking cruises into the West sea.' Thus the account in the Norse saga *Heimskringla*.[7] Another reason for the raids was the acquisition of goods – precious objects that could be ransacked from Christian monasteries and churches and, more importantly, captives for sale in the slave markets of Byzantium and the Arab lands. The term 'viking', derived from the Norse word for a bay, came to mean the perpetrators of pillage rather than the act of pillage itself, reflecting this pursuit of gain.

Once established in the islands these Norse exiles appear to have wasted little time before returning to raid whence they had originated, for the *Historia Norwegiae*, written in about 847, notes that Harald took a great fleet to the northern islands to suppress pirates from the clan of Rognvald who had made the islands their lair and used them as a base for raiding. The same passage in the *Historia Norwegiae* claims that Rognvald's seafaring marauders had destroyed the Picts in the northern islands and taken over not only those lands but, through their assaults, had also brought 'under their rule' Northumbria, Caithness and Dublin.[8] For good measure we are given a potted history of the pre-Norse days of the islands when they had been inhabited by Picts and 'Papae'. The latter were priests of the Celtic church, although the author of the *Historia* mysteriously misinterprets their abandoned books as more typical of 'Africans adhering to Judaism', perhaps because the pagan Norse were reminded of Hebrew and other texts already encountered in their forays to the Black Sea. Of the Picts, they are said to have 'little exceeded pigmies in stature; they did marvels in the morning and in the evening, in building walled towns but at midday they entirely lost all their strength and lurked, through fear, in little underground houses.' These remarks have been dismissed as the coloniser's typically prejudiced dismissal of an inferior race, although they could carry a disguised reference to people taking refuge in brochs or souterrains. The takeover of the northern islands may have been a gradual process but a complete takeover it became, at least in cultural terms. Almost all the old place names that survive in Shetland and Orkney to the present day are Norse in origin. A very rich archaeology testifies to the long centuries of habitation by the Picts and their Celtic forebears but of the naming of this landscape almost nothing has endured beyond echoes in the names of a few islands, for example, Unst and Yell, and Orkney itself, accepted to stem from *orc*, a boar.

What happened to the Pictish inhabitants of the islands? It is unlikely they were all murdered by the Norse. The women and children would have been more valuable as material for the slave markets, not to mention as wives and concubines for the victorious incomers. A few Picts may have escaped to the south as refugees and others would simply have become a resident underclass in a new hegemony. In the archaeological layers excavated at Buckquoy, the occurrence of Pictish artefacts among the deposits of Norse age has been taken to reflect peaceful integration of newcomer and native. In recent years analysis of the DNA of the residents of the northern islands and the mainland has shown the persistence of 'Pictish' genes and in Iceland it has been discovered that, while genes of Norse origin predominate among the men, women have genes of mainly Celtic origin, an inheritance that is taken to mean the Norse seafarers tended to collect their female companions along the way across the Atlantic.[9]

The subjugation of the islands and the Atlantic fringe brought the Norse up against the Celtic heartlands. A victory over the men of Fortriu is recorded for 839 in the Annals of Ulster, in which leading men of both the Picts and Gaels met their deaths 'and others fell, almost without number'. The dead included Wen, son of the Pictish King Onuist, and Aed, king of Dal Riata. Unfortunately, the annals give no location for what must have been for the indigenous peoples a devastating defeat. Hindsight allows us now to see that this was a major turning point in the history of the Highlands as, in the following 60 years or so, the old Pictish kingdoms disappear from the historical record. We have very little information with which to explain this, but a series of military defeats must have made a significant contribution to the rapid decline of Pictish power. In 864, 'the destruction and devastation of Fortriu by the Scandinavians' is noted, with the additional tidings that 'they took away many hostages in pledge of tax. And taxes were given them for a long time afterwards.' The Norse struck at Fortriu from the Irish Sea as well as from the far north. In 866, 'Olaf and Audgish went into Fortriu with the foreigners of Ireland and Scotland, and they raided all the land of the Picts, and took hostages from them'. In 875 the Danes who had by this time taken over Northumbria attacked from the south-east and wrought 'great slaughter' of the Picts possibly at Dollar. We can almost relish the grim glee with which an Irish chronicler noted in 873 the death of the Norse King Godfrey in Ireland 'of a sudden horrible pestilence, for so it pleased God'.

It is during this troubled time that there appears in the historical record a ruler whose name has been taught to every Scottish school bairn. We know him as Kenneth MacAlpin and he has long been identified in popular history as the man who united the Celtic peoples in a single kingdom. In truth we know very little about him. The Viking raids on Iona led to the closure of its scriptorium and, with no monks keeping a record of important events, much in the following decades went unrecorded, except in the barest detail in chronicles distant in space and time from the scene of the happenings that interest us. Accounts of Kenneth's acquisition of power were written or rewritten centuries after his time, and their authors cleaned up and mythologised what must undoubtedly have been a messy struggle for power. Kenneth, or Cinaed mac Ailpín, to give him his name in Gaelic, may have belonged to the Cenel nGabrain or to some other minor kindred in Dal Riata, or he may not have been a Gael at all but a Gaelicised Pict – at the time of his death in 858 the Annals of Ulster calls him king of the Picts.

This melding of two peoples, hitherto often enemies, did not happen without bloodshed. '. . . when Danish pirates had occupied the shores, and with the greatest slaughter had destroyed the Picts who defended their land – Kenneth passed over into, and turned his arms against, the remaining territories of the Picts; and after slaying many, drove [the rest] into flight. And so he was the first of the Scots to obtain the monarchy of the whole of Albania, which is now called Scotia,' states the Chronicle of the Canons of Huntingdon. It goes on, 'In the twelfth year of his reign, he [Kenneth] fought seven times in one day with the Picts, destroying many, and confirmed the kingdom to himself.' The ancient texts may be enigmatic but the overall message is clear. The Picts, or some of the Picts, resisted takeover by the Gaels but were overcome. What may be legend but perhaps with root in fact appears in the writings of Giraldus Cambrensis, the twelfth-century Welsh chronicler. He describes how the Scots invited all the nobles of the Picts to a feast and once their guests were made careless by food and drink pulled the bolts from the benches and slaughtered them before they could struggle from the collapsing furniture.[10] The date of 843 is usually given as the start of Kenneth's reign but struggles to overcome rivals probably lasted for some time in the midst of a period of crisis for the Gaelic realm in the south-west Highlands. The severe pressure from the Norse was forcing the Gaels to abandon their heartland in Kintyre and the islands, displacing them eastward and in effect forging the hitherto separate Gaels and southern Picts into union.

We have almost no information at all as to what was happening around the Moray Firth at this time. One possible scenario is a movement of Gaels to escape the depredations of the Norse along the west coast, a migration

that may have been spasmodic and along different routes. The Great Glen and the Glen Carron–Strath Bran routes are two obvious major ways from west to east, especially for small population groups on the move. It is very possible that the west coast and much of the hill country was sparsely populated but, even so, the incursion of even a relatively low number of refugees into Pictish territories around the Firth must have caused at least considerable tension and likely further social disruption. The first record of Ross occurs in a document ascribed to the eleventh century, a life of Saint Cadroe, in which in a legendary version of the arrival of Celts in the north there is mention of them arriving in Rossia [*sic*] from the west along the river Rosis (now the Blackwater).[11] There are two other aspects to our projected scenario that bear on the future history of the north. The first is that some Gaels stayed behind and integrated with the incomers to create a new hybrid Celto-Norse society. The second is that the Gaels who moved to the Moray Firth area are likely to have belonged to the Cenel Loairn kindred, and may have seen themselves as different from the kindreds who moved more directly eastward into the southern Highlands and spawned the Alpin dynasty. This could have been the roots of the struggle that was

Sueno's Stone in Forres is now enclosed to prevent erosion of the elaborately carved surface. The sequence of scenes with their military figures commemorates some clash of arms but when and between whom remain mysteries. One major possibility is that it marks the defeat of the indigenous Pictish society by the incoming Gaels under Kenneth MacAlpin.

within two centuries to emerge between rival claimants for the throne north and south of the Mounth. Before Kenneth died in 858 in his royal capital at Forteviot, he had brought the relics of Colum Cille to Dunkeld, an act seen as symbolically sealing in place the Gaelic takeover of the Pictish heartland. In the space of some 60 years a separate Pictish kingdom faded into memory. The term Alba begins to appear in the chronicles from around 900, the Gaelic name for the new nation.

During their long expansionist phase, the Norse secured dominion of the northern mainland. The Christian chronicles and annals become even more terse and selective than normal on this topic but we can turn to the equally laconic Norse sagas for a view from the other side. The *Historia Norwegiae* states that Norse pirates in Orkney took over Caithness. The book notes the large whirlpool called the Swelkie in the Pentland Firth, the kind of lore that travellers with seafaring in their blood would have been inclined to talk about. We also have *Orkneyinga Saga*, the only Norse saga primarily devoted to events in the north of Scotland and, although it mingles legend with history, it gives us more detail about the Norse period in our story than we have about many a later time. It tells us that after his scattering of the pirates in the northern islands Harald the Fairhaired gifted these territories to one of his earls, Rognvald. The latter, with Harald's agreement, handed Shetland and Orkney to his brother Sigurd. 'Earl Sigurd became a great ruler,' continues *Orkneyinga Saga*. 'He joined forces with Thorstein the Red . . . and together they conquered the whole of Caithness and a large part of Argyll, Moray and Ross'. The text goes on to say that Sigurd built a stronghold in the south of Moray, possibly a reference to the seizing of Burghead, where charcoal dating to around 865 has been found in the ramparts and is evidence for violent confrontation.[12]

*Orkneyinga Saga* was not written until around 1200, over three centuries after the events that concern us here. Oral transmission of stories can hold tenaciously to some facts while letting others go with carefree abandon. Did the place names Caithness and Moray mean the same in 1200 as they meant in 900? We cannot be certain of our political geography but there are pointers to help us recreate our map. The later chroniclers who recorded the Pictish legends of a northern Scotland divided between the seven sons of Cruithne lump Caithness and Sutherland together as the province of Cait. The present names for the two counties derive from Norse – from Katanes, the headland or promontory of the Kats, and Suðrland, south land – but Cataibh, the Gaelic name for Sutherland, carries in it the older source.

Therefore, for the Norse as well as the Picts, the name element of Kat or Cait may have referred to the whole breadth of the northern mainland as far south as the major topographical features of Strath Oykel and the Dornoch Firth. The precise meaning of Kat in this context eludes us – did it refer to the wildcat having a totemic significance for the inhabitants, or was it as a nickname, or was it a Norse attempt at an indigenous Celtic tribal name?

How Sigurd and his followers overcame local resistance in the north is recounted in macabre and semi-legendary detail in *Orkneyinga Saga*. A certain Maelbrigte is named as the 'earl', to use a Norse term, of the Gaels. Where this leader was based is not mentioned – it could have been somewhere on the south side of the Firth – but it is recounted how a meeting was arranged between him and Sigurd to settle their differences by combat. This is where the legend kicks in, with a familiar motif: 'Each of

The bridge over the Brora river was at one time the only example of its kind in Sutherland and gave rise to the name of the settlement – Norse for 'bridge-river'. The lower stretch of the river is a peaceful place today but the old ice-house on the left of the picture, with the curved turf roof, is a reminder of former activity.

Tarbat Ness with its distinctively banded lighthouse rises like a maritime crossroads in the middle of the Firth. In this view, the hills of Sutherland and Caithness lie to the north.

them was to have forty men but on the appointed day Sigurd decided the Scots [Gaels] weren't to be trusted so he had eighty men mounted on forty horses.' Maelbrigte was not fooled by this hoary trick but honour would not allow him to retreat and in the ensuing clash Sigurd and his followers were unsurprisingly victorious. To show off their triumph they set off for home with the heads of their enemies tied to their saddles. Maelbrigte happened to have protruding buck-teeth and one of these scratched Sigurd's thigh so that the Norse earl fell sick and died. He was buried on the north bank of the Dornoch Firth, supposedly on the site of the farm of Ciderhall, the name a corruption of *Sigurðarhaugr* (Sigurd's mound).

The story of the Norse settlement in the sagas is filled with symbolic colourful incidents such as this. They tell of struggles for territory and power among the Norse themselves along with a few homely touches that appealed to the saga writers' audience. Einar, one of the Norse earls, for example, is credited with the introduction of peat-cutting – at Tarbat Ness of all places.

# CHAPTER 4
# FORGING OF A PROVINCE

The Pictish language itself dwindled and finally disappeared. How this great cultural shift took place is impossible to establish from the surviving chronicles but the indigenous tongue could have been displaced by the incoming one as some key developments tipped the status balance in favour of the latter. The laws of the Gaels seem to have been given precedence over Pictish law; in lines inserted in the Chronicle of Melrose it states that Kenneth was called the first king because 'he first established the Scottish laws, which they call the Laws of Mac-Alpin'.[1] Gaelic was the language of the church and the tongue of the new ruling class from the west. It had prestige. Speaking Pictish would have no longer been an attractive proposition for anyone with ambition. In these circumstances, in an oral culture, a language can fade from use and then from memory in perhaps only three or four generations. How the ordinary Pict, the peasant in his farmstead, fared during this cultural shift can only be guessed at but, although his own head may have been regarded as of little worth, his labour in tilling the land would still have been needed, and it is likely that life went on much as before, with the familiar rhythms repeating as the seasons changed and as his unwritten mother tongue went into decline. Such a pattern of transfer of status on to an incoming language was to be repeated 600 years later when Scots replaced the Norn dialect in the northern islands, and again more recently when English threatened to eclipse Gaelic itself. It is also possible that the Gaelic of the incomers and the Pictish of the indigenous people melded together to some extent, a more likely development if the two tongues were both Celtic. In an oral culture this could have happened without ever being recorded, especially if the literate elite had little regard for the local dialect.

The little we know of Pictish hangs on some terms used in place names. The most common and best known of these is *peta*, meaning a share or

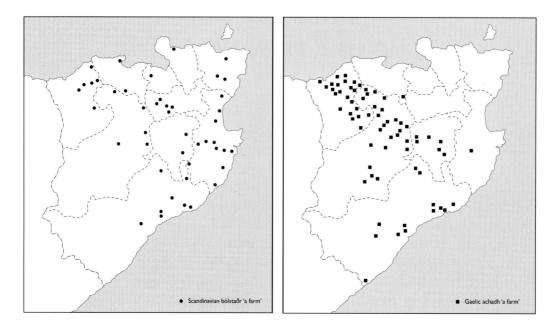

Scandinavian bólstaðr 'a farm'

Gaelic achadh 'a farm'

The maps show the relative distributions of -'ster' and 'ach', two important place name elements in Caithness. The Norse place names are concentrated towards the north-east of the county whereas the Gaelic element forms a distinct band across the south-west. The distribution supports the argument that the Norse settled the fertile north-east before the Gaels coming from the south-west reached the area.

portion of land, and now found as the common place name prefix 'pit' throughout the Pictish heartlands. Around 300 of these names have been listed but some are now impossible to locate.[2] It has been noted, though, that almost all of the 'pit' names have a second element derived from Gaelic, a repeated conjunction that has been interpreted as evidence either for an incoming Gaelic-speaking elite taking over Pictish properties or the adoption by the Gaels of a handy Pictish word for naming their lands. A strong case has been made for the naming process with 'pit' to have taken place between 800 and 1000, just the time when the Gaels were taking over the country.[3] The Book of Deer, an illuminated gospel manuscript surviving from the early 900s, contains records of some land transactions relating to the monastery that give a unique insight into land ownership in this period and provide examples of the use of 'pit', for instance in 'Donald, Giric's son, and Maelbrigte, Cathel's son, gave the farm of the mill – *pett in mulenn* –to Drostan'.[4] Other place-name elements of Pictish origin are *carden*, a thicket, as in Kincardine and Pluscarden; *pert*, as in Perth; *lanerc*, as in Lanark, *pevr*, as in Strathpeffer; and *aber*, a name for the confluence of water courses, equivalent in meaning and usage to the Gaelic *inbhir*, anglicised as inver. Beyond these place-name elements, there are probably more traces of the Pictish language still with us than we realise, hidden within place names.[5] There may also have been a tendency among place-name scholars to seek origins for names in Irish Gaelic simply because knowledge of Pictish has been so limited. The Gaelic place-name scholar W.J. Watson concluded that the name 'Ross' fits well with a derivation from Gaelic *ros*, a promontory, but as the term has cognates in other Celtic tongues – *rhos* in Welsh, *ros* in Breton – it could as easily be a Pictish survival. The name Moray itself first appears in writing in around 970 as *Moreb*, a spelling that leads philologists to see it as *mor-tref*, Brythonic for 'sea home'. As we have seen, Pictish may well have been a Brythonic dialect. Moray also appears as the Celtic form *Muireb*. The variant *Muref* makes its entrance in the late eleventh century in the Annals of Ulster and was also taken into Norse in *Orkneyinga Saga* as *Maerhaefui*. Various meanings have been ascribed to Muref/Moray – from sea or plain. To the Norse the Moray Firth itself was *Breidafjordur*, the wide firth.

That the original province of Moray extended from the Beauly river, referred to as the Forn, a derivative of Farrar, in some sources,[6] to the Spey fits well in the geographical picture we are building. To the east of the Spey lay the territory centred on Banff. This name appears first in the written record in a charter signed by David I in 1124 as *Banef*, in another instance in 1185 as *Banb*, and at last in its current form in 1289. It has been suggested that it related to the word *boyne* from the Gaelic *buinne*, a stream, or it may derive from *ban-ath*, Gaelic for white ford, or *banbha*, Gaelic for sow, but these are not fully convincing. A stronger possibility is that it may come from an Old Irish name for Ireland itself – *Banba*, in this case presumably being a tribute to the Gaels' homeland. The possibility that it may be a derivative of a Pictish word cannot be dismissed. Buchan, historically the district extending to the east of the Deveron river, is possibly derived from the Brythonic or Pictish *buwch*, 'cow', giving a name that means place of cattle and is appropriate still. Some settlement names may have Pictish roots, for example Keith from *ced*, a wood. For Forres a proposed derivation is the Old Gaelic words *for* and *ras*, meaning something like 'under/beside the copse', but could this be a disguised Pictish name? Moy is from *magh*, a Celtic term for a plain. The town of Tain has a name that has long defied satisfactory explanation – it is thought to derive from a pre-Celtic root that means water, a root that lies behind several river names across Europe, but there is a remote and tantalising possibility, or perhaps only wishful thinking, that it is Pictish. The Gaelic for Tain is *Baile Dhubhthaich*, the town of Saint Duthac. Our rivers bear the oldest names we have, stretching back to pre-Indo-European languages, the tongues that were spoken in these islands before the Celts arrived. The Ness and the Nairn, the Spey and the Shin, are all examples of pre-Indo-European names surviving through several changes of language, altering slightly to accord with the sounds of each succeeding tongue. Several of the rivers around the Firth were thought by W.J. Watson to bear witness to a practice familiar from recent history – the naming of natural features by a colonising incomer to remind them of the distant home left behind. Under this proposal, the Findhorn and the Deveron, for example, were held to be 'white Ireland' and 'dark Ireland' named in memory of Erin. The '-earn' in Auldearn would be another example. W.H. Nicolaisen does not accept this and suggests that the final syllable in these cases more likely represents a pre-Celtic, possibly pre-Indo-European root.[7] Elgin is possibly 'little Ireland'. Dornoch is Gaelic, from *dornach*, a stony or pebbly place, fitting enough in view of the glacial moraines and gravels in the district, and Cromarty is from *crom* and *bagh*, crooked bay.

The successors to Kenneth MacAlpin had to struggle for several decades to hold together their territory in the face of the Norse threat. The Moray Firth area fell between the emerging realm of Alba to the south of the Cairngorms and the Norse earldom in the far north, peripheral to both but ignored by neither. No such area of rich land could be and it became a contested zone. In the Chronicle of the Kings of Scotland for the years 889–900, it reads: 'Donald, Constantine's son, held the kingdom for eleven years. The Northmen wasted Pictland in that time.'[8] Walter Bower in his *Scotichronicon* has more to say about this. In his version of events, Donald rejected overtures from the Norse to ally with them in common cause against the Anglo-Saxons on the grounds that he could never help infidel pagans attack fellow Christians. This scruple soon disappeared when the Norse accepted baptism but before Donald could act with the Norse his attention was drawn to 'the regions beyond the Mounth' where, says Bower, 'certain evil brigands' were terrorising the population. Bower often uses the term 'brigand' without elaborating on who they may have been, but they were probably Gaels who were opposed to Donald and in favour of a rival, probably local dynasty. Donald came north to deal with them but 'after a brief illness in the town of Forres he suddenly died'. A suspicion of poison hovers around the king's demise, but it may have been an entirely natural occurrence. Donald was buried on Iona. A few years later, in 904, according to the Chronicon Scotorum, 'Ead, king of Pictland, fell against the two grandsons of Ivar and against Catol with five hundred men'.[9] No location is given for this clash but it is possible Ead was a local ruler in Moray.

The frontier between the new nation of Alba and the lands claimed by the earls of Orkney must have been a very fluid one, a loosely defined any-man's land shifting north or south under short-term political pressure, but with the Dornoch Firth and Strath Oykel roughly marking the boundary. The presence of a handful of Pictish-Gaelic place names north of the Dornoch Firth – Pitgrudy, Pittentrail, Pitarxie, Pitmean, Pitfour, and Kincardine at Ardgay – lead one to speculate that Loch Fleet and Strath Rogart may have been at one time a border zone, although the distribution of these names may be markers for how far north the Gaels had extended their hegemony before the Norse arrived or for the existence at one time of some hardy Gaelic settlers within the Norse area. The scatter of Norse place names in Easter Ross suggests as well that there was, at least from time to time, a significant Norse presence well to the south of Strath Oykel.[10] The most prominent of these names is Dingwall. The derivation is obvious enough – from *thing-vollr*, meaning 'assembly field' – but the history hidden behind the name is far from clear. It could mean there was a Norse settlement at the head of the Cromarty Firth but it could refer to this place being a convenient ground in the frontier zone where Norse and Alban leaders could meet to thrash out diplomatic and political problems.

The kings of Alba had also to worry about enemies to the south. In 934, the king of the Anglo-Saxons, Aethelstan, added his might to the assaults on Constantine II's fledgling realm and despatched a fleet that ranged as far as Caithness. On the death of Constantine in 944, the crown reverted to the direct line of descent from Kenneth MacAlpin, continuing the now-established pattern of switching alternately between the descendants of the MacAlpin brothers, Kenneth and Donald. This state of affairs may not have been acceptable to the rulers in Moray who may have nursed an antipathy towards their southern kin, resenting what they saw as an unjustifiable continual exercise of power by the Alpinid dynasty. Malcolm I, who became king in 944, led his army into Moray and slew Cellach, the local leader and possibly a man threatening the Alpinid grip on the crown. Some mystery surrounds the violent death of Malcolm himself in 954: he was assassinated, but were the perpetrators from Moray or from the Mearns? Walter Bower had no doubts on this score. He presented Malcolm as a king who had the best interests of his whole realm at heart – 'it was his custom in any year in which he was not prevented by other important business to make a progress through the provinces of his kingdom, sitting in judgement over thieves and suppressing the wrongdoing of brigands . . . at last through the conspiracy of certain men . . . by the treachery of the men of Moray he was killed at Ulerin [Blervie, in Moray] . . .'[11]

There is a tradition that the Albans secured a victory over the Norse in the vicinity of Gamrie where in the eighteenth century a patch of heath west of the parish kirk was still called the Bloody Pots and two Norse skulls could still be seen built into the kirk wall. The chronicles mention a victory over the Norse at Cullen in 962. This was, it seems, an ambush set by the Albans near the mouth of the river, and in the conflict Indulf, the successor to Malcolm I, was killed. Bower portrays Indulf as rushing bravely into battle, throwing aside his armour to pursue the fleeing Norse all the faster only to be fatally struck by an arrow from one of the longships. With Indulf's death, the leadership of Alba fell to Malcolm I's son, Dubh, a man, in Bower's words, of 'dove-like simplicity . . . who loved peace and quiet' but still had the capacity to avenge wrongdoing harshly. 'The northern subjects of his kingdom [surely a reference to Moray] were oppressed by the robbers from their own area, although frequently before he [Dubh] had curbed their wickedness by the severity of the law,' says Bower. For robbers, again we should probably read rebels against the Alpinid dynasty. Dubh came north with his army, established his base in Forres, and sent troops out on patrol but his guards became lax and his enemies made their way into the royal chambers at night, seized the king and carried him off, later killing him and hiding the body under a bridge at Kinloss. According to legend, the murder by 'the treacherous nation of Moray', in the words of the Chronicle of Melrose, was such an abomination that the sun did not shine again until the corpse was found. Dubh was buried on the island of Iona in the ground hallowed as the resting place of kings. After his assassination the crown went to a distant cousin, Cullen. It thus stayed within the dynasty of the house of Alpin. If the assassination had been an attempt by the leaders of Moray to seize the throne, it had failed.

In Pictish society it had been the custom for a ruler to name his successor from among his kindred, under the widespread Celtic practice of tanistry. Over time the supreme position could switch between related families. During the tenth century, among the descendants of Kenneth MacAlpin, it became the practice for a king to be succeeded by his nephew or his brother if his own sons were too young or unable to rule. In the

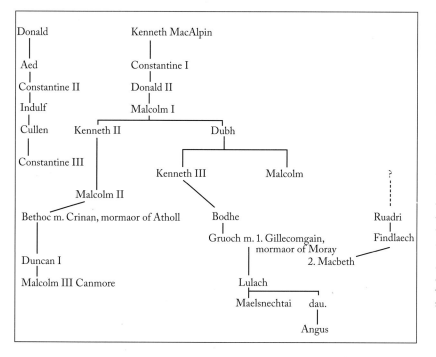

Donald
|
Aed
|
Constantine II
|
Indulf
|
Cullen          Kenneth II
|
Constantine III

Kenneth MacAlpin
|
Constantine I
|
Donald II
|
Malcolm I

Dubh

Malcolm II

Kenneth III          Malcolm          ?

Bethoc m. Crinan, mormaor of Atholl          Bodhe          Ruadri
|                                              |              |
Duncan I                                    Gruoch m. 1. Gillecomgain,   Findlaech
|                                              mormaor of Moray
Malcolm III Canmore                                    2. Macbeth

Lulach

Maelsnechtai     dau.
|
Angus

The Alpinid dynasty and its link with Moray. This diagram of the line of kingship passes through the Alpinid generations and links to the house of Moray. Macbeth's grandfather Ruadri, who was killed in 1020, is described as king of Alba in the Annals of Ulster but as mormaor of Moray in the Annals of Tigernach. The last of the house of Moray was Angus, defeated and killed in 1130. For a detailed examination of the complex politics of this period of history, see Woolf, 2007.

early decades of the eleventh century, during the reign of Malcolm II, the restriction grew tighter, to the offspring of a single line of descent, the line with royal blood, anointed by God. Under this new disposition, in the view from Scone the ruling families in Moray and in other regions of Scotland were now provincial magnates. In the north they were designated by the noble title of *mormaor*, possibly a Pictish term and usually taken to mean 'great steward' – its rendition in Latin as *comes* led to it later becoming equated with earl.

Downgraded they may have been, but the mormaors of Buchan, Moray and Ross were still in the front line in the continuous confrontation with the Norse. By the mid 900s Norse expansion was at last reaching a limit, the stouter opposition on the more densely populated southern shores of the Moray Firth enough to enforce a stalemate. According to tradition, Malcolm II himself enjoyed a great victory over the Norse in 1010 in what is now the parish of Mortlach. Another success over the Norse has also been ascribed to Malcolm II, this time at the small village of Kaim near Duffus; once a pillar marked the place and some said that Kaim took its name from the defeated Norse leader, Camus. This is similar to a legend associated with Corrimony in Glen Urquhart, another place name alleged to derive from a 'Danish prince'. Neither Corrimony nor Kaim are Norse in origin but the stories do testify to the impact the Norse had on the folk memory of their victims. There are also 'forgotten' battle traditions associated with Loch Ashie and Blar nam Feinne, near Altnacardich, in the vicinity of Inverness.

The terse prose of the Norse sagas, legend and history recast as art, recounts the struggle for power in the north between Alba and the earldom of Orkney. At some point during the middle of the tenth century, possibly when Indulf ruled in Alba, Skuli, the younger brother of Ljot Thorfinnsson the earl of Orkney, rebelled against his sibling's rule. The first encounter between the brothers resulted in defeat and flight for Skuli but, re-equipped with men by the Alban king and by Earl Macbeth, Skuli tried again. The resulting battle in 'the Dales of Caithness' was hotly contested and at last the Albans gave way and turned tail, reluctant to die for a Norse would-be earl, leaving the hapless Skuli to fight on until he fell dead. Ljot established himself in Caithness but the Scots, this time led by Earl Macbeth, returned to confront him anew with a large army. The forces met in battle at Skitten, now a bare stretch of high ground a few miles inland from Wick, and once again the Scots were defeated, although Ljot was mortally wounded in the fighting and later died. The reader will probably not have passed over the name Macbeth without a frisson of recognition. This man, however, was not the future king, although he may have been a forebear. The sagas recount another battle on the moor of Skitten during the earlship of Sigurd the Stout, the nephew of Skuli and Ljot, against Finnleik, a Scottish earl. This is probably Findlaech, mormaor of Moray. Sigurd won, helped by fighting under a banner woven by his sorceress-mother.

Sea power played a crucial part in this contest between Norse and Gael. The battles on the southern side of the Firth – at Gamrie, Cullen and elsewhere – were fought within a short distance of the sea. It may seem odd at first glance that the northern encounters were fought at Skitten, so far north that it is within the Norse heartland. The Alban forces could have penetrated Caithness on foot but this would have taken days of marching or riding through empty moorland and it is far more likely that they attacked by sea. The most feasible landing places for a fleet on the Caithness coast are Wick bay or, more likely, the 3-mile sickle of sand in Sinclair's Bay, whence it is only a short hike of 2 or 3 miles to the braes of Skitten. Each side used the Firth to attack the other but neither could maintain mastery of its waters for long.

Findlaech survived the second battle at Skitten but in 1020 he was killed by his nephews in a contest for power within the Moray dynasty itself. One of these nephews, Gillacomgain, met a grisly end in 1032 when, according to the Annals of Tigernach, he was 'burned along with fifty of his men'. The man responsible for this deed was probably Findlaech's son Macbeth who now married Gillacomgain's widow Gruoch and became mormaor of Moray. Two years later, Malcolm II died to be succeeded by his grandson Duncan. Now we feel we are on familiar ground but again it is ground where art and history render the footing less than secure. In Walter Bower's account of Duncan's reign and death, we have a repetition of a familiar scenario. Just as with Malcolm I and Dubh, the good king Duncan was killed through the 'wickedness' of the leaders of Moray. In Duncan's case, Bower says he was fatally wounded in secret by Macbeth at Bothgofnane before being carried off to Elgin where he died. Under the old Celtic custom, Macbeth's claim to kingship may have had some validity but times had changed, and his name was to be blackened by historians – not to mention playwrights – writing to defend the bloodline of the Canmores. As is now well known, Macbeth's reign – from 1040 to 1057 – was far from disastrous. The Chronicle of Melrose noted that 'in his reign there were productive seasons'.[12] This may have been the outcome of several decades of gradual improvement in the climate. In Macbeth's time such fortune would have been ascribed to providence and taken as a sign of God's approval of the new king. Macbeth's pilgrimage to Rome in 1050 where he is said to have scattered money to the poor 'like seed'[13] also suggests that he could safely leave his country for a long period, a journey he would not have been free to make if there existed any serious threat to him at home.

Malcolm, the son of Duncan, was not prepared to leave the Moray lineage in power and, with the aid of the English king and an Anglo-Norse army from Northumberland, defeated Macbeth's forces at Dunsinane in 1054. Macbeth became a fugitive, residing among his Moray supporters for three more years, until his death at Lumphanan in August 1057, whereupon the Moray men and possibly others chose his stepson Lulach as king. Lulach remained a threat to Malcolm until he too was eventually ambushed and killed in the following spring in Strathbogie. Lulach had a son who could have succeeded him, but in 1085 the Annals of Ulster noted that 'Maelsnechtai, Lulach's son, the king of Moray . . . happily ended his life.' The 'happily' suggests that he relinquished secular power for a short time of peace and prayer in a monastery before death claimed him – it was not unknown for leaders to do this.

The house of Canmore, as it came to be called after the nickname of Malcolm III, now ruled all of Alba and was to do so for more than 200 years, but it took long for the leading men in Moray to reconcile themselves to being a provincial rather than a national power. Malcolm, the nemesis of Macbeth, died in 1093, precipitating a struggle for his crown among his siblings and sons that worked to produce a series of short reigns. In 1116, we find a record in the Annals of Ulster of the grandson, Lodmund, of one of the Canmore kings, being killed by the men of Moray. We have no details of this slaying but clearly it was still extremely risky for any important member of the Canmore dynasty to venture north of the Mounth, at least not without considerable military force behind him. And that was about to happen.

Alexander I died in 1124 and was succeeded by his younger brother David. The new king was almost 40 years old and had spent his life in the Anglo-Norman milieu of the English court since the age of eight, whither he and his close relatives had fled from the country after the death of their father Malcolm Canmore. David's elder sister married Henry I of England and he took as his own wife Maud or Matilda, daughter of the earl of Northumbria and a granddaughter of William the Conqueror. A widow ten years older than her new husband, Maud brought to the union two sons from her earlier marriage and extensive estates. David thus acquired in England not only wealth and a strong position within the new Norman dynasty but, more importantly for the future of his own kingdom, new attitudes to kingship and statecraft. Under the leadership of William the Conqueror – or Bastard, depending on one's point of view – England had been submerged by the expanding wave of feudalism, one of the great cultural shifts in European history. Now it was to be Scotland's turn.

David's vision of how a kingdom should be structured meant a feudal arrangement of land ownership and rank by which everyone accorded him the supreme status of overlord and sovereign anointed by God, alongside and bolstered by a universally acknowledged Catholic church with its own hierarchy of clerics. All needs, temporal and spiritual, were thus encompassed in a single system of authority. Not everyone in Scotland shared the royal vision. Almost inevitably, Moray was one centre of dissent.

At the time of David's inauguration as king, the province was ruled by the same Celtic dynasty that had already contested the rule of the Canmore family. The mormaor was now Angus, a grandson of Lulach and therefore a great-grandson of Gruoch, who had been Macbeth's wife. Angus formed an alliance with Malcolm, a possible illegitimate son of Alexander I who felt that he had a good claim to the throne. With 5,000 men at their backs they struck south over the Cairngorms. David was in England. The moment must have seemed opportune. On 16 April 1130, the royalist army under the leadership of David's constable, Edward, confronted the men from the north. The encounter took place at Stracathro about 3 miles north of Brechin. Angus fell in the battle, along with some 4,000 of his followers but the casualties among David's forces were barely a quarter of that, according to the Annals of Ulster. Moray's fate was sealed – it was to become a part of an expanding Scottish realm and its own leaders were henceforth to remain subordinate. It is significant that the Ulster annalist, still imbued with Celtic ideas of kingship, saw Angus as a king whereas to the Norman monk and chronicler, Robert de Torigni, Moray was an earldom, and a rebellious one at that. Thus were Scotland and Europe changing.

# CHAPTER 5
# A NEW ORDER

The crushing of the Moray forces owed much to the knights at David's command, tough adventurers from Normandy and neighbouring parts of northern France and Belgium who, in return for lending their prowess, expected reward. David readily held to custom and granted them lands, as he had been doing since his accession to power. In around 1124, for example, he had given Annandale to Robert de Brus, son of a knight from Brix near Cherbourg who had crossed the English Channel with William the Conqueror. Most of the Normans settled in the southern part of David's kingdom but by 1286, under his successors, they had penetrated up the east coast and around the Moray Firth to Easter Ross. One of the most prominent was Freskin, a Flemish knight, who received shortly after the Stracathro victory the lands around Elgin, good fruitful acres that possibly had been the home territory of the defeated Angus. It was an excellent base from which to continue advancement in the north, as we shall see. Freskin's descendants took the designation *de Moravia*, establishing a dynasty synonymous with the region. That the takeover by Norman and Flemish knights would meet with some resistance might be expected. What is surprising is that there seems to have been so little, or at least this is the impression given by the admittedly scanty surviving evidence. The incoming knights may have been accepted as the new power in the land, but there remains the suspicion that the later chroniclers may have played down and ignored or distorted political turbulence in Moray. On the other hand, the indigenous leaders, depleted and demoralised after the Stracathro defeat, may have made their accommodations with the new regime, perhaps impressed by David's appointment of his nephew William fitzDuncan in place of the fallen mormaor, Angus.

With the conquest, feudalism arrived on the shores of the Firth, with the monarch granting fiefs of lands to his loyal followers, his vassals, in return

for military service. The fiefdom holders, ranked with various titles from earl downwards, granted portions of their land to their tenants in return for services of various kinds. The new lords and knights stamped their authority on their new fiefs by building for themselves fortified strongpoints. These took initially the form of a mound with dwellings surrounded by a palisade, traditionally called a motte and bailey, all built of timber in the first instance but often later reconstructed in stone. Across Europe at this time, the imposition of feudalism represented for indigenous inhabitants an erosion of freedom. Freeholding farmers became rent-paying peasants. Woods and rivers became hunting preserves. The Norse inhabitants of Orkney and Caithness, with their social system of free farmers of roughly equal standing in the earldom, must have watched the Normans settling along the southern shore of the Broad Firth with particular concern. Was it a fear of next to be threatened or was it simply imitation of a new fashion

Duffus Castle, the headquarters of the powerful Freskin family first erected as a motte and bailey in the twelfth century, dominates the plain between Elgin and the coast. Until the nineteenth century the waters of Loch Spynie lapped the castle grounds but now the mound and its ruins dominate fertile fields.

Opposite: Security took precedence over comfort in the siting of medieval castles. This is Buchollie, a stronghold of the Mowat family, on a sea stack on the Caithness coast at Freswick a few miles from John o' Groats. This is also believed to be the site of an earlier fortification called Lambaborg, described in *Orkneyinga Saga* as a base of the Norse chieftain Svein Asleifarsson.

that sparked the building of stone castles such as Oldwick and Buchollie in the far north?

The old Celtic territories of Mar and Buchan, to the east of the province of Moray and straddling the lowland gateway to the Moray Firth area, comprised a single mormaorship in the eleventh century until in about 1113 Alexander I had it divided, making Mar the territory between the Don and the Dee, and Buchan the region north of the Ythan river to the Deveron. The foundation charter for the abbey of Scone names the two earls, using the term *comes* for their rank, as Gartnach and Rothri. Gartnach was succeeded by his son-in-law Colban – a Celtic name – who fought for King William (1143–1214) and called his son Roger – a Norman name. Roger in turn named his son Fergus – a Celtic name – but his granddaughter, who married William Comyn in about 1214, was Marjory or Margaret, Norman again. In this series of personal names we see the fusing of Norman and Celtic cultures.

David secured his rule over Moray on a firmer basis than simply through a military occupation and the planting of loyal followers. As a king he was here to stay. He needed the fruits of peace and an administrative structure to harvest them, and set about installing it through the bishoprics of the church and an institution new to the north of Scotland but already common in Western Europe – the burgh. There was often an intimate connection between the two, between the civil and spiritual powers, and both were instruments of what was in effect a colonisation of the north. A pious man as well as a political realist, the king gave much support to the church. The last entry in the Book of Deer, poignantly marking the end of the Celtic period and the advent of the new feudal Norman–Celtic regime, has the wording: 'David, the king of the Scots, to all his true men, greetings; Know that the clerics of Deer are quit and immune from all service [required] of laymen and from unjust exaction . . . I strictly command that none shall presume to inflict any injury upon them or their cattle'.[1] Written at Aberdeen in the 1130s or early 1140s, the entry is witnessed by three bishops (of Dunkeld, Caithness and Brechin), two earls (of Fife and Angus) and sundry others. Most of the personal names are Gaelic in origin.

David's father, Alexander, had created the bishopric of Moray. The first bishop on record is Gregorius, but it is highly likely that he was never permanently resident in his diocese and may have been based instead in the more central – and safer – location of Dunkeld. A similar appointment of a titular bishop – Andrew – was made by David in respect to Caithness in 1147, more a declaration of a foreign policy interest than a realistic establishment of a new see. From the Norse point of view, the earldom of Orkney, including its mainland portions in Caithness, lay within the diocese of the bishop of Nidaros in Norway. There is nothing in Norse sources to tell us how the locals saw these initiatives of David. Was Andrew an interloper in their eyes, the first move in a chess gambit played by the Scottish king to seize control of the far north? Andrew's name appears as a witness on several royal charters and it seems likely that he never strayed very far from the royal court before his death on 30 December 1184 at Dunfermline. In 1136 David sent a letter to the earl of Orkney to ask that he afford his protection to the monks in Dornoch, a routine request and at the same time a reminder to the Norse of the king's interest in the north. Presumably these monks had arrived recently on the fringes of the Norse earldom. David is known to have stayed at Duffus Castle in the summer of 1150 and inspected the building of the abbey of Kinloss, founded as a Cistercian house. At around the same time he granted to Andrew bishop of Caithness a place called Hoctor conon 'free from all service except army', recorded in a document written at Scone. The location of Hoctor conon is unknown. These charters from the king are all we have to afford us glimpses of the new order, but they present significant information. A charter to the monks of the church of the Holy Trinity of Urquhart written between 1150 and 1153 begins like most with the sonorous Latin address *'omnibus hominibus totius terrae suae Francis Anglicis et Scottis'* – to all men of his land, French, English and Scots. The document goes on to list the properties and other dues for the support of the clerics, including pasturage for the livestock, fishing rights on the Spey, a grant of 20 shillings from the rents of the burghs of Elgin granted first in 1136, and the thain's fishing rights at 'Fochoper'.

The reference to the 'thain' or thane is a reminder that this term first appears during David's reign for an official appointed by the king to rule a district called a thanage. It may have been simply a new name for an existing system of administration. The word itself has its roots in the Old Norse and/or Anglo-Saxon term *thegn*. An alternative term for thanage was shire, another word of Anglo-Saxon origin, but to avoid confusion with the more familiar definition of the latter as the area under the judicial authority of a sheriff, it is preferable to stick to thanage. The thanes themselves were intermediate in social status between nobles and free peasants, and one of their main functions was to collect revenues due to the royal house.

Thanages first appear in the Lowlands and spread through Strathmore and up the east coast, skirting the Cairngorms, to Buchan and Moray, almost as if they were springing up in the wake of the advance of royal hegemony. In the Moray Firth area, we find them in a broken belt between Aberdeen and the mouth of the Deveron, a second grouping to the west along the fertile Moray lowlands and one lonely outlier, Dingwall, beyond the Black Isle. Some of the thanages had names that are still in use now – Boyne, Conwath [Conveth], Aberkirdor [Aberchirder], Nathdole [Natherdale] in Banffshire; Moy, Brothyn [Brodie] and Dyke in Moray.

The distribution of thanages, the broken sequence with large intervening or adjacent tracts belonging to none of them, raises a few questions. If these were the only thanages, should we see them as areas particularly suited to the agriculture of the time, and therefore the most productive and most desirable; or were they the areas where incomers loyal to David were placed in charge? Thanages would have been highly valued as land within the gift of the monarch, and also served the practical purpose of being places where the king could be sure of hospitality as he moved around his realm. In an age when the preservation of food was difficult, it was the custom for a monarch to keep an itinerant court, staying in different places for varying

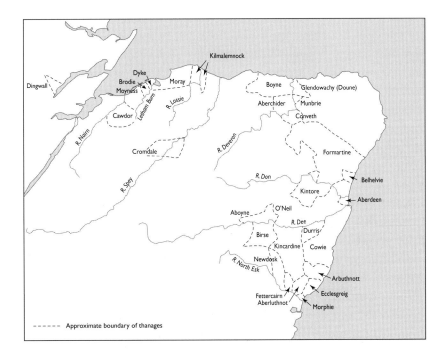

Thanages in the North-east. These divisions of land may have existed before the reign of David I (1124–53) but it is during his time that they first appear in the record. The thane in charge of a thanage was a royal official responsible for administration and the collection of dues. Dingwall is the only known thanage north of the Great Glen.

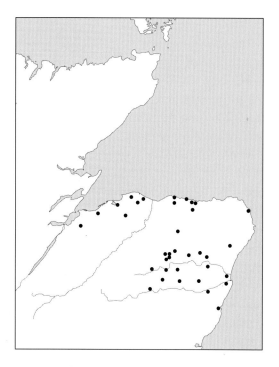

lengths of time. In this way, rents and dues paid in kind could be consumed.

With the organisation of thanages came a reordering of judicial power in the form of sheriffdoms. The first to be established in the Moray Firth area were those of Aberdeen, which included Buchan, and Banff, which date from around 1136. Their boundaries corresponded almost exactly with the Gaelic provinces that preceded them. A sheriffdom of Moray may have been formed at the same time but this may have been a parchment exercise. Probably the impossibility of one jurisdiction coping with such a province stretching west through rough country became fairly quickly apparent and eventually it was subdivided. A lingering propensity to rebel against southern domination may also have encouraged a heavier input of authority.

The motte, a natural or man-made mound surmounted with or surrounded by defensive structures forming the 'bailey', is the characteristic form of military fortification associated with the introduction of feudalism to the north. Mottes could be thrown up in a short space of time, perhaps a few weeks, to serve as a base, and their distribution on the southern side of the Firth reflects the westward penetration of lowland Scots and Norman-French into hostile Moray territory during the twelfth century.

Separate sheriffdoms appeared by 1232, centred on Elgin, Forres and Nairn, and by 1266 on Cromarty, leaving the rump of the province always under the sheriff of Inverness. The jurisdictions also became divided as sheriffs who held land in different districts brought all their own territory under their own jurisdiction, so that, for example, the sheriff of Nairn could end up with authority over patches of ground that had originally been under the sheriff of Elgin. A relic of this shifting pattern existed until 1975 in the name of the county of Ross and Cromarty, originally a patchwork of two sheriffdoms.

David was succeeded in 1153 by his 12-year-old grandson, Malcolm. The first year of the young lad's reign was marred by a great famine and an outbreak of a 'pestilence among animals'. Malcolm continued the policies of his grandfather in promoting the church and the burghs, as shown by several charters that survive from his reign. On Christmas Day 1160, he granted to Berowald the Fleming two areas of land between the Lossie and the Spey and a 'full toft' – *toftum plenarium* – in the burgh of Elgin, the income from which would pay for the service of a knight based in Elgin castle. Berowald assumed a title from his new estate, low-lying and

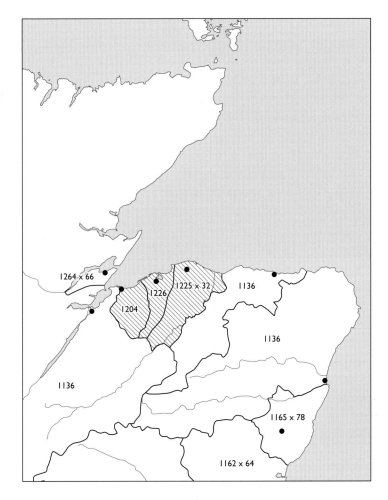

The judicial division of the sheriffdom was introduced to the north-east and the Firth area during the reign of David I (1124–1153.) In this map the dates of founding show a progression from east to west. A sheriffdom of Moray was probably established by 1150 before being subdivided into smaller jurisdictions centred on Elgin, Forres and Nairn. The sheriffdom of Cromarty, formed by 1266, initially included only the area around the town. A second phase of sheriffdom formation took place much later, with the creation of the shires of Ross and Caithness in 1503, and Sutherland in 1633. In the later Middle Ages, the office of sheriff came to be occupied by successive members of the same family but these so-called hereditable jurisdictions were abolished in 1748.

virtually an island, and referred to himself as *de Insula*. In time, probably in fairly short order, the Gaelic word for such land – *innis* – became the family name, Innes. Most of the surviving charters relate to the church. Malcolm was of a particularly pious frame of mind – he was nicknamed 'the maiden' – and he was generous to the spiritual power abroad as well as at home. The Hospital in Jerusalem received the income from a toft in each of his royal burghs, and three silver merks from the shipping income in Perth went to the Abbey of Tiron in France each year. A half of all the oil from whales washed ashore between Forth and Tay was granted to the monks of Dunfermline to light their altars. Kinloss Abbey received the site of a mill on the Altyre Burn, a toft in Elgin, Forres and Inverness, and patches of land at Burgie and elsewhere.

The Chronicle of Holyrood notes for the year 1163 that Malcolm 'transported the men of Moray'.[2] This enigmatic comment would seem to describe a wholesale banishment, though of which men and how many will never be resolved. There is a tradition in the history of the clan Mackay that they are descended from forebears driven from Moray to the north-west of Sutherland, a barren and inhospitable place deemed suitably remote to serve as a place of exile for a recalcitrant and untrustworthy enemy. Another aspect of this tradition relates the transplanted Mackay ancestors to the Forbes kindred in Aberdeenshire but the origin traditions of the Forbeses add little to this notion and settle more soberly for tracing their lineage from Duncan de Forbeys, granted a charter for the family lands in 1271–72 by Alexander III.[3] Another tradition links the Mackays to the clan Morgan. The only mention of this mysterious kindred is in the dedication of land to the church made by Colban of Buchan in around 1135, recorded in the Book of Deer, where it is stated that Donnachac, son of Sithig, toisech of Clan Morguinn, acted jointly with the earl.

Malcolm was succeeded on the throne by his younger brother William in 1165, an altogether more belligerent and campaigning monarch who was later nicknamed 'the Lion'. Among the charters that survive from his reign is one containing the earliest mention of a burgess of Inverness – Gaufridus Blundus in the Latin, or Geoffrey Blount. William also granted to the bishop of Moray a mill on the Lossie and the right to take fuel from the royal forests around Elgin, Forres and Inverness as long as these extractions did not infringe the rights of the burgesses. This could refer to timber but it is also highly likely that it included peat.

That there should have been resistance to Malcolm's policy of banishment is hardly surprising and it may well have been these aggrieved, ousted men who found in a certain Donald Ban MacWilliam a leader who promised to restore them to what they had lost. Donald Ban was the great-grandson of Malcolm Canmore and his first wife, Ingebjorg. His grandfather had been king, Duncan II, briefly in 1094 before the crown had reverted to the offspring of Malcolm's second wife, the saintly Margaret, and had stayed there. Donald Ban, therefore, had reason to challenge for power and, in the glens of Ross and Sutherland, found men ready to fight at his back. 'Supported by the treason of some treacherous men [MacWilliam] had first wrested from his king by the importunity of his tyranny the whole of Ross; and afterwards holding for no short time the whole of Moray, he had occupied, with slaying and burning, the greater part of the kingdom –

aspiring to the whole . . .'[4] The peasants of Moray receive no consideration in the chronicles of the period beyond the cliché 'slaying and burning'.

William reacted in the way of a fighting king. In 1179, he led troops into the area of Ross, '. . . along with the earls and barons of the land . . . with a great and strong army,' according to the Chronicle of Melrose, 'and there they strengthened two castles . . .' The two castles were Dunscath, in the lee of the North Sutor of Cromarty, and Ederdover, also given as Eddradower, at Redcastle on the Black Isle. Obvious declarations of power, these strongholds were strategically sited to receive supplies by sea and to overawe any man bent on challenging the Canmore dynasty. Donald Ban may have retired into the fastnesses of Ross to avoid battle during this first campaign by William but, according to Fordun, he 'continued in his accustomed wickedness', forcing William to mount a second expedition – this time in 1187 – to bring the intransigent would-be usurper to heel. Donald Ban withdrew into Ross again. While William was encamped in Inverness, his earls 'sent their men to plunder'. The Chronicle of Melrose goes on: 'And they found MacWilliam with his followers upon the moor that is called Mam-garvia near [*prope*] Moray'. No one knows this location: it is clearly a Latin version of the Gaelic *mam garbh*, meaning a rough, rounded hill, a simple enough description that would fit any number of sites around the head of the Beauly Firth and the description of being 'near Moray' is hardly useful. MacWilliam's fate was now sealed – '. . . presently they fought with him [MacWilliam] and by God's help slew him, with many others . . . Blessed be God who has betrayed the wicked.'[5]

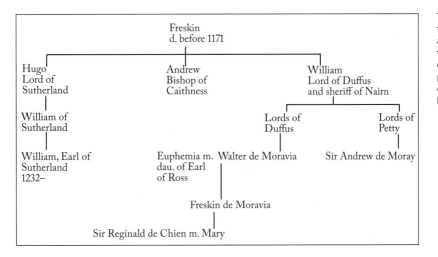

The Freskin dynasty founded in Moray by a Flemish knight in the service of David I eventually achieved prominent positions on both sides of the Firth.

In its version of this victory, the Chronicle of Holyrood adds: 'And peace, long disturbed, was restored to the king and the kingdom through God's mercy and virtue.' Sadly for the ordinary people, the peace was not restored for long. Ross may have fallen quiet but beyond it lay the Norse-dominated province of Caithness and Sutherland. The earldom was at this time in the hands of Harald Maddadarson, a man in his sixties, of mixed Celtic–Norse descent and a capable survivor. His father Maddad had been earl of Atholl and a nephew of Malcolm Canmore, and his mother Margaret a daughter of an earl of Orkney. After first sharing the earldom with Rognvald Kali Kolsson, the builder of Saint Magnus Cathedral in Kirkwall, Harald became sole earl after Rognvald was assassinated on a hunting trip in Caithness. *Orkneyinga Saga* tells us that his second wife was a daughter of 'Earl Malcolm of Moray'. This must be none other than Malcolm MacHeth whom Malcolm IV had made earl of Ross in around 1162; from the distance of Iceland where *Orkneyinga Saga* was probably written, niceties of Scots political geography could easily have been overlooked. With Moray and Ross secure, William shifted his sights further north to the Orkney earldom and in the 1190s backed an attempt by Harald Ungi, a grandson of Rognvald Kali Kolsson, to grab half of Caithness from the older, incumbent Harald, no doubt intending that it would fall thereafter under his own influence. But the rising failed. Earl Harald overcame and killed his rival of the same name in a battle near Thurso, and laid all Caithness under his rule, says the *Saga*, a state of affairs that naturally did not please William. He tried another move to get his hands on Caithness, this time enlisting the aid of Rognvald Godrodarson, the Norse 'king' in the Western Isles and reputed to be the greatest fighting man 'in all the western lands'.[6] Rognvald invaded Caithness with a strong force – which Harald coolly avoided by staying in Orkney. After some weeks Rognvald felt the need to return to the Hebrides and left Caithness under the administration of three stewards. Now Harald decided that he had to wait no longer and crossed the Pentland Firth to land at Scrabster. John, the second bishop of Caithness, the successor of the Andrew appointed by David I, had his residence at Scrabster and hurried to meet the earl apparently intending to welcome the newcomers and possibly offer some conciliatory explanation of their cooperation with the departed Rognvald. Harald was having none of it – this bishop in any case was an appointee of the Scots dynasty, and he had already annoyed the earl by refusing to collect from the inhabited households of Caithness one penny per year for a donation Harald had

pledged to the papacy. The Orcadian earl had the bishop tortured, blinding him and cutting out his tongue. John survived this ordeal – the *Saga* notes that he had his sight and speech miraculously restored by Saint Tredwell – and died in 1213, outliving his tormentor by several years. The stewards fled south to the safety of William's court while Harald violently restored his authority in Caithness.

William now came north in person at the head of an army, 'a great force from every part of the land' according to *Orkneyinga Saga*, exploiting the attack on a bishop of the church as reason to invade, perhaps convinced that he could no longer waste time trying to acquire this northern province through proxies but had to tend to it himself. Harald gathered his own forces but he was outnumbered, unable to attack the Gaels who filled the whole valley of Eysteinsdalr on the Caithness border. Messengers came with the welcome news that William was willing to reach a settlement. Peace came at a hefty price – all the Caithness men except those who had stayed loyal to the stewards of Rognvald had to hand over a quarter of their revenues. The Chronicle of Melrose records a further twist to the story. In 1197, it says, Harald's sons, Thorfinn and Roderic, met the king's vassals in battle near the castle of Inverness and were routed. Roderic was killed but presumably Thorfinn escaped with his life. 'In all things blessed be God, who has betrayed the wicked,' observed the chronicle.[7] According to the

The valley of Ousdale near the Sutherland–Caithness border – this is the *Eysteinsdalr* of *Orkneyinga Saga* where King William and Earl Harald Maddadarson negotiated a settlement in around 1195.

Chronicle, Harald himself was taken south by William to be kept as a hostage in the castle at Roxburgh – further from Kirkwall could hardly be possible within Scotland – until Thorfinn decided to surrender. *Orkneyinga Saga* makes no mention of Harald being a hostage at Roxburgh but adds the detail that Thorfinn was blinded before the war was over. To help mend relations with William and the church, Harald bound himself and his heirs to make an annual grant of one mark of silver to the canons of Scone. When he expired at last in 1206, he was succeeded by two of his younger sons, Jon and David, who jointly ruled until David's death seven years later.

Ross and beyond continued to be a troublesome region for the Canmores. William's forces – the old lion himself was now in his 69th year and probably stayed at home – had to return to deal with a rising in 1211, led this time by Godfrey, a son of Donald Ban MacWilliam, who had returned from exile in Ireland in an attempt to avenge his father. The rebel was captured, brought before William's son Alexander at Kincardine, decapitated and hung up by the feet. In the late summer of 1214, the aged king made his last expedition to Moray. The earl of Orkney, now Jon Haraldsson in solitary rule, sailed south to pay suitable homage and left his daughter in the royal household as a token of his loyal intentions. Shortly afterwards when he was back in the south, William fell ill and survived for only weeks before he died. He was succeeded by his son, who became Alexander II.

Commanding the 1211 expedition to Ross had been William Comyn, justiciar of Scotia, the name then given to the entire region north of the Forth before it was adopted as the Latin for the whole kingdom of Scotland. For years the justiciar's family had been on the rise. Originally de Comyn, the name has long been associated with the town of Comines near Lille, but more recently it has been suggested the family sprang from among the clerks working in the cathedral cities of north-west France.[8] If this is so, such a relatively humble background may have been a spur to their ambition, and proof that in the long run brains could be more profitable than brawn. The first Comyn to make a mark in Scotland was also a William, a William who became David I's chancellor from about 1136 and who had learned his business in the chancery of Henry I of England. Chancellor William's nephew Richard acquired lands in Northumberland and found a wife from a junior branch of the Canmore kindred, a grand-niece of Malcolm Canmore himself. By the time of his death in about 1179, Richard had extended his land holdings into the Scottish Border country. It was his son who was now loyally serving the king and had been rewarded with the earldom of Buchan after marrying into the family that already held it. The Comyns may have been typical of some incoming Norman-French in showing an apparent keen willingness to become adopted in their new country and happy to rewrite history to make good their credentials. There is one story that none other than the great Charlemagne himself came to Inverlochy in the Great Glen to visit a Pictish king and sign a treaty that was witnessed by 16 members of the Comyn tribe; the survival and possibly the creation of this myth no doubt owes much to the Comyns themselves.[9] As happened in some instances in Ireland, the Normans contributed to a new hybrid Gaelic aristocracy. For status, Margaret of Buchan could hardly have done better than marry a Comyn.

Alexander II found his wives abroad, first a daughter of King John of England and after her death in 1238 a daughter of the powerful Picardy knight, Enguerrand de Coucy. Thus was the Scottish royal dynasty linking itself to other European houses. Warfare and diplomacy were also directed outward. Bricius, bishop of Moray, went on a mission to Rome in 1215, and Alexander involved himself in English affairs. Yet all was still not secure within Alexander's extensive backyard. His father had achieved an understanding with the Orkney earls over the governance of the northern mainland in which it seems to have been accepted that the Scots king had a certain interest but no more, the territory remaining essentially a Norse dominion. The extent of that interest was now to be tested, and tested by the church in the person of the third bishop of Caithness to be appointed under the Scots. This man was called Adam. Of uncertain status at birth (he was left as an infant by an unknown mother at a church door), he clearly had abilities and ambition and rose through the clerical ranks to become abbot of Melrose before he was appointed to the Caithness see in 1214. In 1218 he went with the bishops of Glasgow and Moray on pilgrimage to Rome. On his return to Caithness, he abandoned the humility of a pilgrim for more worldly desires, increasing the taxes laid on the bondsmen, the free farmers, who supplied goods and services to his residence at Halkirk in the dale of the Thurso river. For every 20 cows in his possession, the Caithness bondsman customarily surrendered to the bishop a span of butter – about 5½ kilos – each year. Adam increased this demand to every 15 cows and then to every ten, in effect a doubling of taxation. The bondsmen carried their protests to the earl but when he chose to do nothing attacked the bishop's residence, broke in, killed a monk and roasted Adam to death. The shock of this outrage against a bishop reverberated through Europe –

it was noted in Iceland and in Rome – and Alexander acted quickly and forcefully to re-establish lawful authority. Coming north with his soldiers, he apprehended the late bishop's parishioners and had the hands and feet hacked from 80 of them. The earl of Caithness and Orkney who chose not to intervene to save the hapless bishop was Jon, the only surviving son of Harald Maddadarson. In 1231, he met a grisly end himself, stabbed in a cellar in Thurso in what was apparently a local vendetta. He was the last of the Norse earls. Alexander granted the earldom of Caithness to Magnus, a son of the earl of Angus, and divided from it the lands of Sutherland, forming the latter into a new earldom that he gave to William of the Freskin kindred in Moray.

In 1230, there was another rising in the Ross–Moray area by those the Chronicle of Lanercost called 'certain wicked men of the race of Mac-William' whose leaders it names as MacWilliam, his son and someone called Roderic.[10] The Chronicle provides almost no detail of this challenge to Alexander beyond stating 'by the vengeance of God they and their accomplices were betrayed and . . . successfully overcome'. Alexander ordered MacWilliam's infant daughter to be murdered, by having her head smashed against a stone pillar in Forfar, a gruesome execution carried out in public as a demonstration of the king's power but against the wishes of his clergy.

For all its strategic importance as a political unit lying athwart the neck of northern Scotland like a buffer between Norse and Scots-Gaelic territory, the early history of the earldom of Ross is obscure. The first firm record of the title comes when Alexander II granted it, before Christmas 1225, to Fearchar Mac-an-t-Sagairt in reward for loyal service. Several historians of the clan Ross have pushed the existence of the earldom a century further back, to a dynasty of Celtic mormaors with the name of O'Beolain associated with the lay abbots of the Applecross monastic foundation. One account has it that Fearchar brought rebel heads to lay before the royal feet and was given a knighthood for his trouble.[11] Alexander was determined to add the western seaboard to his realm. As the writer of one Icelandic saga has it, he was 'covetous of dominion in the Hebrides and constantly sent men to Norway to demand the purchase of the lands'. What could not be bought could be seized and it was on an expedition to secure the Hebrides that the king died suddenly of natural causes on Kerrera in July 1249. He was succeeded by his seven-year-old son, who was crowned Alexander III at Scone a week later.

# CHAPTER 6
# BURGHS AND FIEFDOMS

The burgh was an idea transplanted from Europe that served the interests of the monarch, essentially a means to encourage trade and provide the king with a steady source of income that did not depend on the vagaries of the growing season. In David I's time, the royal household gained its entire cash revenue from burgesses.[1] The privileges extended to the burghs also benefited the townspeople, the burgesses, in that they were granted protection, at least on parchment. The first royal burghs were created in the Lowlands, at such places as Perth and Stirling, Dunfermline, Peebles and Linlithgow, and a few nobles followed suit by setting up their own small burghs in the lee of their castles. Their appearance in Moray was a deliberate act of colonisation. The Celts, whether Pict or Gael, and the Norse showed no great inclination to live too close to their neighbours. The largest settlement they appeared able to thole was probably akin to the later fermtoun, a household with its extended family and collection of retainers and slaves sprawled around it. Presumably, they also had a network of hamlets and monastic communities that served as foci for markets, administration and social or religious life. The area of Inverness at the junction of major routes and Burghead on its defensible headland are obvious candidates for such places, and Portmahomack and its monastic centre may have been another. The new type of burgh was different, a settling together of strangers, of families unrelated by blood, a huddling for commerce and protection in a concentrated area with the potential for expansion. A suitable site was divided into plots of ground of a regular size, often by Flemish surveyors, and each plot was assigned to a merchant or a craftsman and his family to encourage them to settle together. A rood, a quarter of an acre, became the standard plot and it was normal to offer it rent free for a fixed period, often one year, to the would-be burgess who in that time was expected to erect a house and establish a business. In

Dingwall, perhaps because it was dangerously near the northern edge of the realm and perhaps also because it was reckoned not to enjoy such good economic prospects, the rent-free period was set at ten years.[2]

The first burghs in the Moray Firth area were at Banff, Elgin and Forres. At the mouth of the Deveron, Banff was ideally placed for communication by sea and acted as gatekeeper on a major route inland. It had a castle of some kind from an early date and Malcolm IV was residing there in 1163. The positions of Elgin and Forres were probably determined by the same criteria – proximity to the sea, which penetrated closer at the time of their founding than it does now, and access to routes leading inland. On high ground on the south side of the Lossie, Elgin was probably already a political centre for the Gaelic province, the obvious place for David to choose as his royal burgh and capital. To the north, across the grey sheet of Loch Spynie, David's loyal follower Freskin raised his own stronghold of Duffus Castle. Forres also began on high ground, a ridge of shingle and sand with Cluny Hill and its crowning castle at the west end. The oldest reference to Elgin is a grant by David I of 20 shillings per year from the burgh to the monks of Urquhart for their vestments, and the Urquhart monks were also beneficiaries of the earliest document to mention Banff, a charter signed there also by David. Neither is dated. One reason for the absence of documentation, as if one were really needed for an age when little was put down in ink, is given in a much later charter, by James IV in 1496, to Forres in which it is written '. . . clearly understanding that the charters, evidents and ancient writs of the foundation of infeftment of our Burgh of Forres . . . have, through the devastation of war, by fire and other means been destroyed, burnt and annulled . . .'[3] At the time that James renewed the Forres charter, the town had suffered greatly from the actions of 'lawless and wicked men', and its commerce had been severely reduced 'to their [the burgesses] great detriment . . . and to our great loss and damage, with regard to the established customs and dues . . .' The protection granted to a burgh by a royal charter often remained an unrealised ideal.

It may have taken many years for the promise of the burghs to appear – Forres did not start to grow until the reign of David's son, William, and Banff too may have had a slow start. There is considerable uncertainty over the beginnings of several burghs – something that led to some exaggeratedly hopeful claims being put forward by loyal townsfolk in later ages – but Cullen, Nairn and Inverness probably came formally into existence as burghs in the second half of the twelfth century. A charter by

William possibly in 1179, granting to the burgesses of Inverness freedom from tolls and customs throughout the realm and forbidding anyone to trade there without their permission, is the first official recognition of Inverness as a self-governing community. The concept of the burgh spread slowly northwards, with the emergence of Dingwall and Cromarty between 1214 and 1314. Rosemarkie acquired a charter in the thirteenth century as the seat of the bishopric, but was later in 1455 under James II united with the neighbouring Chanonry or Fortrose to form a new burgh.[4] Not all the embryonic burghs survived. Eren, where William had set up a fortress, was attacked and destroyed, possibly by Donald Ban MacWilliam and his followers. The would-be burgh remained a village, Auldearn, its moment in history to come several centuries later. William chose to establish a new base at the mouth of the Nairn but his 'castle' there succumbed to the sea; according to the description of Nairn in the *Statistical Account of Scotland* in the 1790s, its vestiges were still visible at extreme low tide.[5]

Who were the inhabitants of the early burghs? Some of them must have been locals, Gaelic-speaking, perhaps exponents of crafts such as blacksmithing and carpentry and welcome in their communities, but many, perhaps most, of those who became the new class of burgesses were incomers, settlers from the south of Scotland or from Continental Europe. Knights and clerks came across from Flanders and Normandy, and there is no reason to suppose merchants, bravely willing to try the opportunities on offer in the Scottish king's expanding realm, would have been far behind them. The privileges granted to the inhabitants of the burghs extended only to the ruling sections of these communities, the coteries of merchants and craftsmen who comprised the burgesses and freemen. Their servants, the tradespeople and the unskilled labourers who no doubt comprised the bulk of the urban population were legally 'unfree' and had no official say in how the burgh was governed. Apart from the occurrence of a handful of names in a few charters, there is no firm evidence as to the inhabitants of the new burghs until almost a century after their founding, and we have to wait longer still before we encounter surviving burgh records with more detailed information on daily life in the streets and wynds. At Elgin in May 1196 or 1197, King William granted to the Inverness burgess Gaufridus [Geoffrey] Blundus and his heirs the right never to have to engage in trial by combat. Geoffrey bears a Norman Christian name, his surname Blundus is possibly a distinguishing nickname from the colour of his hair, and he may have been Flemish or of Flemish descent. A family called Wiseman –

unfortunately the name gives little clue to the origins of its bearers – was prominent in Moray in the mid 1200s: a Thomas Wiseman was sheriff and provost of Elgin in 1248, a William Wiseman, possibly his brother or son, was sheriff of Forres in 1264 and the owner of the lands of Mulben, and an Alexander Wiseman was sheriff of Forres and Nairn in 1305. The first sheriff whose name we know for Elgin is Alexander Douglas, in 1226.

In August 1261 an enquiry was held into who had possession of some property in Elgin. The surviving record lists the leading men of the district who pronounced a resolution to the problem of who should continue to enjoy the use of what was called the 'king's garden'.[6] Thus we know of 'Ewynus thayne of Rothen, Dugal thayne of Molen, Thomas Wisman prepositus of Elgin, Andrew of Innerlochtyn [Innerlochty], James of Brenath, Hugh Heroc, four burgesses – Richard Brun, Hostyn Grouzbacheler, Robert Diker and Andrew Wysy, Walter of Always [Alves] and Andrew son of Leuyn'. This is a fascinating mix of names – some Celtic, some English/ Scots, some Norman-French. The list also shows people identified by the place where they lived, by their father's name, possibly by a nickname or occupation. The matter of the king's garden affords us more detail on burgh life. It was in the hands of a man called Robert Spinc who had acquired it through his wife Margaret whose forebears had cultivated it to provide the king with its 'fruits', including 'potherbs and mallows' whenever he stayed at the castle. If the king should choose to keep his gyrfalcon there, Margaret's people were paid a penny a day for feeding the 'ostorius', twopence a day for bird food, and a chalder of oatmeal a year. The good men of Elgin decided that Robert and Margaret could continue in possession of the garden. Some of the same men gathered five years later in Inverness to provide evidence for possession in another case, this time whether or not the ancestors of Ewan or Eugene thane of Ratthen, presumably the same man as Ewynnus thayne of Rothen in the previous case, held the land of Mefth from the king in heritage and was therefore entitled to hold it himself. All the jurors testified that King William had given the land and a house, as well as a net on the Spey, to one Yothre MacGilhys in heritage 'for service of one sergeant and being in the Scottish army', and that Ewan was a descendant of Yothre and no reason could be given why he should not possess the same property.

The burghs were the places where one would have been most likely to encounter another of David I's innovations, seemingly humdrum but with far-reaching consequences – a Scottish coinage. No longer need the trading of goods be in the form of barter or payment in kind. A regular, lawful, convenient currency must have boosted commerce, although perhaps only in limited circles in the early years. It is a question as to why it had not appeared before, as the Anglo-Saxons and the Norse had long dealt with coin, yet somehow their Celtic neighbours had not accepted the practice. When David introduced his first coins is disputed, but it may have been in 1136 when his forces captured Carlisle and the mint sited there. The David I penny, the only denomination, was small, weighing 22.5 grains of silver, and had a crude image of the royal head with the words *Davit Rex* stamped on one side, and a cross on the other. Several variants of the basic design appeared, all very rare now.

For the first two centuries or so after the founding of the burghs, we catch only a few glimpses of the trade carried on around the Moray Firth. Early in 1304 four merchants from Saint Omer raised a complaint about goods they had bought in Moray that had been sequestrated, they claimed, by the bishop of Moray and taken to Berwick. The validity of their case is of less interest than this proof that French merchants were trading in the Firth. The goods described are all raw commodities – 32 sacks of wool, value 60 shillings each; three sacks of lambskins, worth two marks each; 36 dickers of hides, worth 11 shillings each; a pisa and half of lard 'in a little pipe', worth 20 shillings; and 'packells' of deerskins, rotten and torn by rats.[7] The oldest customs records for exports from Inverness are for the years 1337 and 1359, when the commodities were also wool and hides. Trade and commerce likely operated on two levels – the local and the distant, with the former much the more significant activity and mediated through markets as well as through individual deals. Timber was another raw material that was abundant around the inner Firth. Famously, a ship was built in Inverness in 1249 for Hugh de Châtillon, Count of Saint Pol, the only indication we have that there were skilled shipwrights, possibly Flemish, in the area at that time.

One of William's charters, dating from around 1180, confirms an earlier charter from his father in the sonorous Latin of the royal chancellery '. . . *hac Carta mea Cofirmasse Burgensibus meis de Aberdoen – Omnibus Burgensibus de Morauia – Omnibus Burgensibus meis ex aquilonali parte de Muneh manentibus liberum ansum suum Tenendum ubi voluerint – quando voluerint . . .*' – 'This charter of mine confirms to my burgesses in Aberdeen, and in Moray, and in all parts north of the Mounth their free hanse to be held where and when they wish'.[8] A hanse or hansa was a company, guild or

The cathedral in Dornoch, now the parish church, gracefully wears its years. It was probably founded in 1225–30 as the episcopal seat of Gilbert de Moravia after he was appointed bishop of Caithness and Sutherland. Sacked in 1570 by marauding Sinclairs and Mackays, and later severely damaged by storms, it has endured several bouts of repair over succeeding years to achieve its present attractive condition.

union, usually of merchants. The word derives from Old High German and Gothic and is most frequently used in relation to the trading organisation, the Hanseatic League, founded in the German cities of the Baltic. Another hanse of cloth-producing towns existed in Flanders and the Rhineland. Evidence is lacking that the Firth burghs were linked in any formal sense, rather they were part of the network of mercantile centres that fed goods into the Continental circuits, possibly through Aberdeen as some kind of entrepôt for the whole north. One commodity for which there was a continual demand in Christian Europe was fish. On the Caithness coast in the vicinity of Duncansby Head, archaeological investigation of extensive middens of fish bones point to the existence in the twelfth century of a fishery producing dried cod, saithe, ling and other white fish on a scale far in excess of what one might expect for domestic consumption. Similar operations could have existed elsewhere around the Firth and may well have been linked to the Continental markets.

Life beyond the burgh boundaries probably continued much as it had done except in the districts where feudalism had a strong impact. The term *dabach* or more commonly *davoch* is a word derived from the Old Irish *dabhach*, meaning a large vat or tub, that was adopted as the measure for the amount of grain a piece of land could be expected to yield and therefore became a unit of land assessment. The area of a davoch varied considerably, depending on the natural productivity of the soil. There may have been a Pictish equivalent or perhaps the custom of assessing land in this way, as being so many davochs, was introduced with the Gaels. The Norse-occupied areas of Caithness, Sutherland, the Hebrides and other parts of Scotland have a unit of land assessment called the pennyland. The two systems overlap but not to a great extent in the two northern counties. South of the Forth land was assessed in terms of ploughgates. In modern terms, a davoch could range from 48 to 72 acres, and a pennyland from 12 to 18 acres, though one should not look for exact area equivalence here. That *davoch* and pennyland continued in use through the medieval period and up until the nineteenth century suggests that the incoming feudal lords took over the existing divisions of the land, and this in turn leads one to conclude they took over the tenants and the agricultural practices of livestock rearing and cultivation as well. In England the feudal barons showed great interest in improving their lands to increase their income and in time the rearing of sheep and the export of wool became a major source of wealth. A similar development occurred in the Scottish Border country but was there anything of this sort around the Firth? The economic histories of the period tend to focus on the Continent's western heartland, the zone from Flanders and northern France through to the Rhineland, with London as a western outlier. The twelfth century saw a surge of intellectual and cultural activity in this western heartland and in the Mediterranean – this was the time of the wandering scholars, troubadours, the growth of religious orders and the beginnings of humanism – but how much of this surge washed up on the Moray Firth shore, a region far from being isolated, must await further exploration.

Four bishoprics – Aberdeen, Moray, Ross and Caithness – fringed the Firth in the thirteenth century. That of Ross could bear the honour of being the oldest, possibly dating from the time of Colum Cille, although the earliest recorded incumbent is possibly MacBethad of Rosemarkie, active in the years 1127–31. The first record for Aberdeen – an earlier see was based at Mortlach – also dates from around the same time where the bishop is named Nechtan. The Caithness diocese is the youngest of the four. Gilbert de Moravia of the Freskin kindred succeeded the murdered Adam as bishop of Caithness in 1222, clearly resolved to place the church on a more secure footing in the far north. He made his seat at Dornoch, in the territory of his cousin, William the first earl of Sutherland, ensured the cathedral there was completed, well staffed and provided for, and created the distribution of parishes in Caithness and Sutherland that has persisted more or less into the present. Another kinsman was Andrew de Moravia, bishop of Moray, and Gilbert followed him in adopting Lincoln cathedral as a model for the episcopal constitution. Andrew's immediate predecessor as bishop of Moray, Bricius, had chosen to have his cathedral at Spynie –

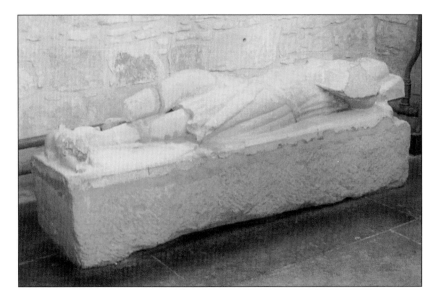

This piece of broken monumental sculpture in Dornoch cathedral is believed to be an effigy of the bishop's brother, Sir Richard de Moravia, who lost his life in around 1259 in a struggle with invading Norse at Embo. In the same clash, the earl of Sutherland is supposed to have seized a horse's leg to use as a weapon when he lost his sword in the thick of the fray, hence the horseshoe on the town's arms. The cathedral became widely known in December 2000 as the place Madonna and Guy Ritchie chose for their wedding service.

it had previously been variously at Birnie, Spynie or Kinnedar since the founding of the see in the reign of Alexander I – but Andrew decided to shift it to Elgin where building began in 1224.

Parish churches and chapels were probably simply built with plain stone walls and thatched roofs, with more sophisticated architecture under Continental influence appearing only in the grander ecclesiastical buildings. Construction of cathedrals began at Dornoch and Fortrose in the early 1200s. The Valliscaulian priory at Beauly dates from around 1230, the Premonstratensian monastery at Fearn from around 1225, and from 1238 on its newer site at Hill of Fearn. Of the Dominican convent founded in Inverness in the reign of Alexander II only a solitary, sorry-looking pillar remains in a gloomy graveyard behind the telephone exchange. Another Valliscaulian house founded in 1230 at Pluscarden may never have been finished before it was destroyed by the Wolf of Badenoch in 1390. It later became a Benedictine house, and the present abbey is a restoration. These religious buildings all show the characteristics of the Gothic style deployed throughout Europe, the church using its wealth to declare its membership of a Europe-wide civilisation.

When they were not engaged in campaigning against each other, kings and lords in the Middle Ages directed their attentions to hunting, a pastime that not only handily kept their horsemanship and weapon skills up to scratch but also provided fresh meat for the table. Hunting reserves, termed forests although not necessarily wooded, covered some extensive stretches of territory on the southern shore of the Firth and appear in charters. For example, in 1226 Alexander II granted to the bishop of Ross the forest of Ineurculane [Invercullen], and in 1291 the bishop of Caithness received 40 oaks from Darnaway for the roof of his cathedral.[9] The imposition of laws to restrict the exploitation of the woods and hills was another way in which the peasantry lost some personal freedom under the feudal system.

During the thirteenth century, in the reigns of the two Alexanders, we come upon the earliest appearances in the surviving evidence of several of the great names that dominate the history of the central and northern Highlands. Among them were the Bysets or Bissets who

The elegant gable of the priory in Beauly, founded as a Valliscaulian house in around 1230.

came to Scotland as loyal followers of King William in the 1170s – the name is of Norman origin from the Old French *bis*, meaning a rock dove, although there was probably little about the Bisset knights that was dove-like. They were granted the lordship of the Aird, the attractive swathe of country between the Ness and the Beauly rivers. John Byset is credited with the founding of the Valliscaulian priory at Beauly. At a tournament in Haddington in 1242, Walter Byset, lord of Aboyne, was beaten in a contest by the young earl of Atholl, a member of the Comyn family. When the latter was found dead in his house soon afterwards, Walter was accused, perhaps unjustly, and banished from the kingdom, along with his nephew, John, the lord of Beauly. The Bissets later regained position – a Walter Buset of Lessendrum was sheriff-substitute in Banff in 1364 – but for a time they lost their standing and all their lands. According to James Fraser in the Wardlaw Manuscript, seven of their baronies went to the Comyns and the Campbells but the estate of the Aird, the lordship of Loveth or Lovat, and other lands in the region were gifted by the king to Sir Simon Fraser, a younger son of Lord Tweeddale, who married a Bisset in 1249. Like the Bissets, the Frasers could trace their origins back to Norman France – one

explanation of the name derives it from *fraise*, strawberry, the plant chosen by the family as a badge of identity. One branch of the family acquired lands in Buchan and became the Frasers of Philorth.

Hakon of Norway, responding to appeals from the Hebrides for protection against raiders from mainland Scotland, made an effort to establish his authority in the Western Isles in 1263 but after an indecisive encounter with the Scots at Largs retreated to Kirkwall where he died in mid December to be succeeded by his son Magnus VI. By this time, as we have seen, Caithness and Sutherland had been split by Alexander II into two earldoms, with Sutherland securely under the rule of the Freskins and within the realm of Scotland. The situation in the earldom of Caithness, however, remained less clear cut. Alexander II had granted the Caithness earldom in 1232 to Magnus, a son of the earl of Angus, who was probably also confirmed in the earldom of Orkney by the Norwegian king before dropping out of the historical record until his death was noted in 1239.[10] A great-grandson, also called Magnus, of the first Angus earl held the Orkney earldom in 1263 and had command of a ship in Hakon's fleet on the last, fateful expedition. While in

Orkney before venturing down the west coast, Hakon or more probably Magnus sent men to extract money from Caithness, contributions the earl probably saw as rightly due to him from his tenants. Unfortunately for the Caithness people, who now found themselves trapped between two contesting powers, Alexander had already taken hostages there to ensure they remained loyal to him. The accounts of the sheriff of Inverness for 1263 refer to the expenses of keeping the Caithness hostages – a penny each per day for 25 weeks. After the defeat and death of Hakon, Alexander sent

The three coats of arms on the wall of Spynie Palace are of the royal house, the Hepburn family and the Stewarts.

troops into Caithness to extract more from the locals. In the autumn of 1264 some Scots who had returned to Caithness for taxes were attacked and relieved of their takings by a Hebridean-Norse warlord called Dugald, son of Ruadri. At last in 1265 diplomatic initiatives between Norway and Scotland began to bear fruit, leading to the signing of the Treaty of Perth in 1266. This agreement finally ceded the Hebrides to Scotland in return for financial compensation, ended hostilities and settled the bounds of the Scottish realm for another two centuries until the acquisition of Orkney and Shetland in 1468. The whole Moray Firth area now fell under the Scottish crown but the region was still marked by ethnic boundaries. Norse remained the daily language in the north-east of Caithness for several more generations, one element in a varied verbal stew that also included Scots English and declining Norman-French in the burghs, Latin in the churches, and Gaelic over most of the region.

# CHAPTER 7
# OCCUPATION

It is the most famous riding accident in the history of Scotland. On a wild night of wind and sleet in March 1286, while he was hurrying homeward to his young French wife, Alexander III and his horse tumbled over the cliffs near Kinghorn in Fife. His untimely death launched long years of conflict, a struggle all the more bitter coming after what was recognised as a successful reign, in retrospect even a 'golden age'. It had been a long reign too – only the middle-aged and the old could remember another king – as Alexander had yet to see his eighth birthday when he had inherited the crown on his father's death in the summer of 1249. At his inauguration at Scone his line of descent had been declaimed by a Gaelic bard. His own mother had been French (she had died only in 1284). More than ever before one could argue that the monarch represented a weaving together of disparate ethnic strands into a single national fabric. To what extent did a sense of nationhood and national identity exist throughout Scotland at this time? The peasants probably identified themselves with community and feudal overlord, the burgesses with their fellow townsfolk, the clanspeople with their chieftains but an appreciation of a more comprehensive identity, shared membership in a political nation, was probably only present in the upper echelons of society. There are signs of this appearing in Alexander's time. The phrase 'community of the realm' is used in documents of the period.[1] Many Scottish lords had to juggle a special dilemma: they held fiefs on both sides of the Tweed and had to distinguish carefully between fealty pledged to the English monarch for one lot of possessions and to the Scottish monarch for the other. These concepts and their practical implications became suddenly crucial with the death of Alexander III.

With his first wife, Margaret, the eldest daughter of Henry III, Alexander had had three children. All of them had died before their father and in 1286 Alexander's only surviving offspring was a two-year-old granddaughter,

Margaret of Norway. This infant was acknowledged as the rightful heir by the Scots nobles and, once she had grown a little, in the autumn of 1290, a ship set off to bring her to her new realm. Messengers rode north to Caithness to meet her, staying en route at Skelbo, Helmsdale and a place referred to as 'Hospital', leaving us with the only clue to the existence of such an establishment on the coast, possibly at Latheron, before riding on to Wick where they received the stunning news that their journey was in vain. Margaret had turned ill on the voyage, the ship had put into Orkney, but the little princess had not recovered. The body of the girl who has been known ever since as the Maid of Norway was taken back to Bergen for interment.

Her death left Scotland without any clear royal heir but with a slew of 13 contending claimants. Among them were John Comyn of Badenoch, John Balliol and Robert Bruce. An observer familiar with the rivalries common among medieval lords may have expected the kingdom to fall apart in the contest for the crown. It says much that, at least for a long time, it did not. Instead the leading nobles and clergy met to negotiate a way out of their difficulty, appointed interim 'guardians of the realm' to govern the country, another indication of a sense of nationhood, and, failing to reach a satisfactory outcome to their deliberations on the succession, asked Edward I of England to arbitrate. Edward chose John Balliol not because he had the strongest claim but because Edward thought he could be intimidated. Balliol is usually seen today as a weak nonentity, exactly as his derisive nickname 'toom tabard' would have him. His three years on the throne were not, however, entirely a wasted interlude. He continued the nation-building of his predecessor, setting in motion measures to create new sheriffdoms in the western Highlands and islands that in the end came to naught but showed clear and sensible intentions, and in the end he defied the bullying of Edward I. The English king responded with bloody aggression in 1296, sacking the town of Berwick and massacring its inhabitants, and defeating the Scots army at Dunbar.

Leaving the shattered Lowlands behind him, Edward led his army – 5,000 horsemen and 30,000 foot soldiers – on a triumphant progress north through Scotland. On 6 July in Forfar he received homage from Hugh de Moray, Hugh Urry and Sir Andrew de Breton, the first in a series of submissions made in Strathmore all the way from Perth to Montrose. On 7 July John Balliol threw himself humiliatingly on Edward's mercy – at Stracathro no less, where 166 years before David I's troops had smashed the power of Moray. Balliol was despatched into England and, after three years as a prisoner, he returned to his ancestral lands in Normandy where he died in 1313. At Montrose on 13 July, Sir John de Moravia, and the two John Comyns, of Buchan and of Badenoch, pledged their fealty to the English king. They were followed in the succeeding days by other northern magnates, including John de Strivelyn of Moray, Reginald de Cheyne or Chien senior and Sir William de Moravia. 'With kingly courage, into the region of the unstable inhabitants of Moray whither you will not find in the ancient records that anyone had penetrated since Arthur' went Edward, noted the Lanercost Chronicle with a fine conceit.[2] The Plantagenet triumph proceeded through Aberdeen, Fyvie, Turriff, Banff, Cullen and on to the west. On 27 July the 'burgesses and community' of Elgin and one Alan de Morref declared their loyalty and over the next three days another 20 knights and prominent individuals did likewise before Edward turned south. When the king was back in Berwick at the end of August the Scots nobility or their representatives were summoned to place their seals on a document of fealty since known in Scotland as the Ragman's Roll. Among the fringe of heraldic insignia are the seals of several Moray Firth nobles – the Badenoch and Buchan Comyns, Sir Andrew de Moray of Petty, the earl of Caithness and the earl of Ross.[3] A large number of the leading men of the kingdom were shipped off to incarceration in England, including the earl of Ross and Sir Andrew de Moray senior who were imprisoned in the Tower of London, Andrew de Moray junior in Chester, John, the son of Alexander de Moray, in Bristol and Reginald de Cheyne junior in Kenilworth.

The unity displayed by the Scots nobility immediately after Alexander's death no longer held. The nobles and churchmen pondered on where their best interests lay – in acknowledging the overlordship claimed so violently by Edward or in retaining loyalty to a more uncertain Scots kingdom, now in effect headless. The north found itself dealing with the classic dilemma of a country under foreign military occupation – whether to resist or to collaborate, not an easy choice when the castles around the Firth, at Inverness, Nairn, Forres, Glenurquhart, Dingwall, Cromarty and Banff, were in the hands of garrisons loyal to Edward and such prominent families as the Comyns and the de Cheynes also supported the new regime. The Comyns had a firm grip on the country from Buchan to the Great Glen, with control of several castles, including Balvenie at the strategic conjunction of several major routes through the hills in the east of Moray,

Lochindorb and Inverlochy. Henry de Cheyne was bishop of Aberdeen, and Sir Reginald held the wardenship of Moray as well as lands in Caithness.

Edward's dominance of Scotland did not remain unchallenged for long. At some point during the winter of 1296–97, the young Andrew de Moray escaped from Chester and made his way home to the Black Isle where he found an ally in an Inverness merchant Alexander Pilche. This Pilche must have been a prominent burgess. The surname – it crops up on several occasions in the scarce records of the period – suggests the family may have been involved in the fur trade, a typical export from the north at this time. In May 1297, Andrew and Pilche led an attack on Sir Reginald de Cheyne as he was travelling to Urquhart Castle from Inverness, and the knight and some of his companions were taken prisoner. On the following day, Andrew and Pilche's party laid siege to Urquhart Castle. This outbreak alarmed Euphemia, countess of Ross, who despatched a messenger to the castle to let the commander know that this attack was nothing of her doing. For some reason, she advised the garrison to surrender – perhaps she was torn by fear for her husband's fate in the Tower of London – but this they refused to do. The son of the countess and his men managed to boost the castle's provisions. A night assault by the besiegers was beaten off and, perhaps accepting that Urquhart was for the moment too strong to be taken, Andrew and Pilche withdrew. In a letter written by the bishop of Aberdeen in Inverness early in August, the clergyman praised the countess of Ross for her conduct and begged Edward to let her husband return. At around the same time, the constable of Urquhart reported on the active zeal of 'John de Laarde', John of the Aird, to whom he was indebted for his and his children's safety, and appealed for John's son, Cristinus, a prisoner in Corffe Castle, to be released and sent north. Cristinus's return, said the constable, would have the effect of winning the country to Edward's cause and gaining the English king favour with the inhabitants. Other Scots prisoners were set free at this time, including some Badenoch Comyns. The men on the spot clearly thought that in 1297 there was still time to win the struggle for hearts and minds. On 24 July, the bishop of Aberdeen, the earl of Buchan and the son of the earl of Mar reported to Edward the suppression of the insurrection in Moray and confirmed the assistance of the countess of Ross. Soon, though, Andrew de Moray made his way south and met up with William Wallace. Their joint leadership secured a tremendous victory over the English forces at Stirling Bridge on 13 September, but Andrew did not survive the fighting; a document of November 1300 refers to him as 'slain at Stirling against the King'.[4] Wallace continued to lead the drive for independence and the restoration of Balliol's reign until the defeat of his forces at Falkirk in 1298 left Scotland once again rudderless and uncertain.

An uneasy stability settled over the Moray Firth lands for a while. Edward revisited Moray in the autumn of 1303 and from Kinloss in September he ordered at last the release of William earl of Ross and his safe escort north. The earl and his retinue which included a cook, servants and men-at-arms took 18 days to travel from London to Berwick. In December William had a new suit of armour made in Dunfermline. This loyalty to the Plantagenet regime secured William his earldom but his new role proved expensive, as we find him informing the English administration in 1304 that he had spent over £1,000 of his own money in trying to suppress the 'foreign isles' and their 'cheventeyns', presumably meaning the Hebrides and their chieftains.[5] Edward had already provided William with the rentals from the lands of Dingwall and Ferintosh but this was clearly not enough to cover the expenses of policing such a vast area. By the following year William was appearing in the English sources as the warden of the country beyond the Spey. All in all, the earl of Ross seemed to be doing well under the new dispensation. In contrast we do not know how successful the dean of Elgin Cathedral was with his plea in September 1303 for timber to repair his house which had been burnt by Edward's army. The danger of insurrection persisted. The earl of Athol advised Edward not to give Aboyne Castle to Sir Alexander Comyn as the country around was savage and full of evil-doers and, in any case, Sir Alexander already held Urquhart and Tarradale; Athol also writes about someone called Lachlan or Lochlan in Ross whose intentions were unclear but who had ordered each davoch of land under his control to furnish a galley of 20 oars, an apparent preparation for campaigning that echoes the Gaelic world of Dal Riata and suggests the old practices were still in operation in the west.

In the autumn of 1305 Edward divided Scotland into four parts and appointed a pair of justices to each one in an attempt to settle the country. For the region 'beyond the Mountains', i.e. the Moray Firth area, these officials were Sir Reynaud le Chien (Sir Reginald de Cheyne – his name appears in several variants in the records) and Sir John de Vaux of Northumberland. The territory was further divided into a series of jurisdictions: Sir William de Berkeley became sheriff in Banff, William Wyseman in Elgin, Alexander Wyseman in Forres and Nairn, Sir John de Stirling in Inverness and Sir William de Mohaut in Cromarty. Neither the king of England nor the king of Scotland had so far managed to push much of a legal presence further north than that. Documents in the English state papers relate to the collection of revenues, the so-called compotus of the king's lands, around the Moray Firth at this time. The earl of Ross sent in the 'issues' of the bishopric of Caithness on 24 June 1305, a sum amounting to £40, for the year 1304–05. From Inverness, through the agency of a member of the Pilche family, came the 'farms' [rents] of the town (24s), the same from the burgesses of Elgin (46s 8d), from the sheriffdom of Nairn (10 marks), from the burgesses of Nairn (50s), the sheriffdom of Forres (£6), the 'burgh farm' for Forres (26s 8d), the 'farms' of Elgin (100s), from the sheriffdom of Banff for the issues of the bailliary (£21) and from the same place for the farms of the sheriffdom (£9 11s 6d). The papers also include the costs of furnishing the clerks who rode from burgh to burgh with an escort to collect these sums, an expensive, lengthy but necessary business that recognised that in the Moray area, as in other parts of the country, lurked men 'who had not yet fully come to the King's peace'. It cost 20s to pay for 20 men to escort the clerks from Aberdeen to Banff in May 1304. Twenty foot soldiers went with the clerks from Banff to Elgin, where they were guarded during their 14-day stay, and both mounted men and foot soldiers from Sir Reginald de Cheyne escorted them on to Inverness 'and there staying with them on account of the imminent peril of enemies, and escort back to Elgin' (cost 60s).

Such was the state of the country in February 1306 when Robert Bruce and John Comyn of Badenoch, the guardian of Scotland in the exiled Balliol's absence, met in the Greyfriars kirk in Dumfries. Bruce had already decided that the moment had come to make his bid for the throne but Balliol was still the lawful king and the Comyns were loyal to him. Exactly what passed between the two noblemen, already with a history of rivalry between them, will never be known but their discussion turned into violence. Bruce stabbed Comyn and his companions completed the killing. Six weeks later Bruce was crowned at Scone. During the first months of his rising, things went against him and he had to flee, spending the winter as a fugitive, probably for part of the time at least on Rathlin Island off the Ulster coast, before returning to resume action in Carrick, his home territory, in February 1307. Now his fortunes began to change. His defeat of the main English force in Scotland at Loudoun Hill in May sent a thrill of excitement through the ordinary folk of the realm and caused many of the Scots nobles to reconsider a man they had previously seen as a usurper.

'. . . I have heard from Reginald Cheyne, Duncan of Frendraught [sheriff of Banff] and Gilbert of Glencarnie, who keep the peace beyond the Mounth and on this side, that if Bruce can get away in this direction or towards the parts of Ross he will find all the people all ready at his will . . .', wrote a pro-English lord in Forfar, shortly after Loudoun Hill.[6]

Early in July, Edward I died, a stroke of luck for Bruce as the new English king, Edward II, was a much less forbidding foe. The enmity of the Comyns, more implacable now since the murder of John of Badenoch, was a more immediate threat. A war for independence had, therefore, to begin with a civil war, and a civil war that had to be fought in Comyn territory to ensure victory. Late in September 1307, Bruce led his army through the southern Highlands to attack the Comyn stronghold at Inverlochy. After the fall of the castle, he struck up the Great Glen to destroy the castle of Urquhart and Inverness. Nairn was burnt. David bishop of Moray and Thomas bishop of Ross supported Bruce, but William earl of Ross found himself isolated and fearing for his life. In November he wrote to tell Edward II what course he had followed, how he had taken up a defensive line with 3,000 men 'at our own expense' on the border of the earldom, possibly along the Beauly river, and pleaded with the English king to come to his aid in the spring. William says that 'good men, both clergy and others' – probably the bishops of Moray and Ross – had managed to secure a truce between the earl and Bruce to last until Whitsun 1308. But we would not have done that, protested William, if Reginald de Cheyne, as warden of Moray, had not been absent from the country. William was not facing Bruce entirely alone but his neighbour, the earl of Sutherland, although anti-Bruce, was not much support and, as for Magnus, earl of Caithness, he was too far north to be much in the picture. The bishop of Caithness was pro-English but the Caithness earl's position at this time remains, in fact, unknown, although he later came into Bruce's peace. Ignoring Ross for the time being, Bruce attacked Elgin and then Banff, his main aim to overwhelm the Comyns of Buchan. John Comyn and his allies gathered their forces to meet Bruce in the field but, as winter tightened its grip on the north-east, the new king withdrew into a defensive lager at Slioch near Huntly. Bruce fell sick and his 700 or so followers feared for his life. In this time of weakness, though, Comyn and his allies failed to press home an attack. Bruce recovered and went once again on the offensive, attacking the Comyn castle at Balvenie and then sacking the Cheyne castle at Duffus. From the environs of Elgin, the king launched an assault to the west, capturing Tarradale, forcing the hapless earl of Ross once more onto the back foot. William Wiseman, loyal to Bruce, led a party to capture Skelbo castle, but a second attempt by Bruce to seize the castle at Elgin on Palm Sunday, 7 April, had to be abandoned when pro-Comyn forces intervened. The final, crucial encounter with Comyn of Buchan happened near Inverurie, on an unknown date but probably Ascension Day, 23 May. Bruce was victorious and sealed his triumph by ordering his troops to pillage the Comyn earldom, slaughtering Comyn loyalists, tearing down farms, burning grain, in effect a calculated act to destroy the power base of his chief foe. According to Barbour, the 'herschip' [harrying] of Buchan was remembered 50 years later.

Those who had not committed themselves to Bruce's banner now thought again. The most significant leader to recognise the inevitable was William earl of Ross. On the last day of October 1308, he submitted at the castle at Auldearn and Bruce pardoned him and received him into his peace. In his verse epic composed in the 1370s on the wars of independence, John Barbour, archdeacon of Aberdeen, was to write:

> The king than till his pes has tane
> the north cuntre, that humylly
> obeysit till his senyhory.
> Swa that benorth the Month war nane
> That thai ne war his men ilkane.[7]

Thereafter, the focus of the struggle shifted to the south. Less than 200 years before, the Moray region had been a hotbed of hostility towards the royal house of Canmore. Now it had become and was to stay loyal to Bruce and to the succeeding Stewart dynasty. In 1312 Bruce rewarded one of his leading followers, Thomas Randolph, with the earldom of Moray, lands that had been in the crown's possession since 1130.

# CHAPTER 8
# HIGHLAND AND LOWLAND

Banff's long history is evident in its architectural heritage, as shown on this plaque beside the entrance to the old kirkyard.

In the later Middle Ages the burghs on the Moray Firth remained not much more than large villages, with gub in the wynds, middens in the street and the smoke from peat fires hanging around the thatched roofs. But unprepossessing as they were, they had potential, pretension and significance far beyond their size. To some extent they had become islands of privilege outside the feudal hierarchies of power that prevailed in the surrounding countryside. Royal burghs were in theory answerable only to the monarch, although this was a guarantee with a health warning – local power was always likely to prevail over that of the usually distant king, at least in the short term, and royal burghs were at times little better off than the burghs of barony that remained subordinate to the church or the magnate who had founded them. When Sir Thomas Randolph was given the earldom of Moray in 1312, burgesses were instructed by the king to obey and assist him but they retained their own rights and liberties under the crown. Thomas was granted the so-called 'great customs', those fees and tolls deriving from foreign trade, but the burgesses kept the 'petty customs', from the burgh markets and inland commerce. The burghs ran their affairs according to a sett or constitution that was common to them all and persisted with little change for centuries. Under the sett, the burgesses and the most prominent tradespeople, the freemen, governed their community. Becoming a burgess conveyed membership of the local elite and licence to participate in trade, but it also required the ownership or renting of a plot of ground, at least one rood (quarter-acre) in size, payment of a burgess fee and no doubt some expenditure on food and drink to ease one's way up the social ladder. It also brought responsibilities, expressed in the Inverness burgh records in 1546 as 'to scott, lot, walk and ward' – to pay taxes and share in the policing and defence of the burgh, the latter no easy task at times.[1] When Nicol Duff was admitted as a freeman and burgess in Forres in October 1585, he took an oath that included an almost identical formula: '. . . sall be leill and trew to ye town and sall scott latt walk and warde'.[2] All the freemen were obliged to attend 'head courts' convened by the provost three times a year, at Michaelmas, Christmas and Easter, to debate and agree the major direction of the local administration. The burgh council, chosen by election among the burgesses once a year at Michaelmas, met more often to attend to routine administration, and the provost and the senior councillors, the bailies, collectively formed the magistracy to dispense justice in the burgh court for local crimes and misdemeanours. In the early days of the burghs, the praepositus had been the officer responsible for the collection of the annual rents and fees due to the crown; the role evolved into that of provost, and the collection of revenue in the 1300s into a system whereby the annual sum due to the crown became fixed and any surplus remained in the hands of the burgesses, a system in places formalised under a charter of feu ferme that allowed the surplus to go to a common-good fund for the benefit of the burgh. Inverness received a feu ferme charter in 1370, Banff in 1372. Unfortunately any early records of burgh administration have fallen victim to time. Those, for example, for Forres survive only from the early 1500s, for Elgin burgh court from 1540 and for the burgh council from 1636, for Banff burgh court from 1650 and for the council from 1674, for Inverness burgh court from 1556, for Dingwall from 1708, and for Wick from 1660. Before these dates we need to turn to charters and other documents for insights into burgh life.

With a small population, it was inevitable that some families came to dominate local affairs. Wisemans were prominent in Moray for a long time, and later Dunbars came to the fore – there were five of that name on the burgh council of Forres in 1562, with four of the name of Urquhart, examples surpassed by the later instance in Inverness in 1673 when 19 of the 21 members of the burgh council were related in some way

This steep lane, Strait Path, in Banff is one of several that link the modern High Street with the old heart of the town on Low Street.

to the provost.[3] The external trade of the burgh rested in the hands of burgess families, but local commerce was much more diverse with stallholders, craftspeople, peasants and fishers practising a variety of trades and selling their wares and services in weekly markets. These were supplemented by larger fairs, usually associated with particular saints and running for several days. Large-scale trade was concerned mainly with such bulky goods as hides, skins, wool and wool fells. As mentioned before, Inverness makes its first appearance in the Exchequer Rolls for customs on exports in 1337 and thereafter there is no further entry until 1359 when the sum of £22 15s for 68 sacks, 6 stone of wool and 49s for hides is recorded as the payment for Inverness, Forres and Nairn together. Regular renders of customs revenue from Inverness do not begin until 1369. The average rendering of customs to the Scottish chamberlain from Elgin for the four years from 1367 to 1370 was £119 5s 6d, mainly for the export of wool, skins and hides. The sailing vessels that plied the Firth in the Middle Ages and linked its harbours with other parts of Britain and the Continent were typically cogs, an example of which was dug from the mud of the Weser estuary near Bremen in 1962.[4] Originally built in or after 1380, the Bremen cog is decked and single-masted, a little over 76 feet long overall with a beam of 25 feet and a cargo capacity of 80 tonnes.

'Be it known to all men through these present letters.' The charter begins with the customary formula, except that this time the words are in Scots – 'Be it knawyn til al men' – and not in the usual church Latin. It is dated 23 July 1393, the location Elgin. Thomas Dunbar, earl of Moray, may have chosen to cast his latest legislation in Scots because the men most directly affected, the burgesses, now used it as their everyday tongue. The earl recorded that he had granted to the 'alderman' [provost], the bailies and burgesses 'al ye wol ye clathe and al uthir thyngis yt gais be schipe owte of wre hafine of Spee vncustomyt' – all wool, cloth and other things that go by ship out of the haven of Spey uncustomed. 'Wyt al men [let all men know] we wil nocht thole' such a practice, stated the earl before adding his seal to the document.[5] The burgesses had no doubt made abundantly clear to the earl their views on this intolerable shipping of goods from the river mouth where whoever was responsible was obviously trying to avoid paying the proper customs charges to the proper authority, themselves, who had by their royal burgh status the monopoly to seaborne trade in the district. The target of the legislation may well have been the bishop of Moray, Alexander Bur, who had already become embroiled in disputes over commerce. The bishop regarded the port of Spynie as the church's property but the earl and the burgesses also claimed they could use it. In 1383 the bishop had arrested a small ship that had entered the haven of Spynie with the permission of the burgesses of Elgin. Neither the burgesses nor the earl had any such right, declared the bishop before he made his arrest. Later that day, when he found two burgesses, named as Philip Bisset and Henry Porter, unloading a cargo of barrels of beer, tallow and flour, the bishop arrested them as well.

Such local disputes were less of a threat to the burghs than the turbulence of national politics. When Robert Bruce died in 1329 and was succeeded by his son, David II, only five years old, with the earl of Moray as his

'Be it knawyn til al men . . .' The earl of Moray's charter to the burgesses of Elgin in July 1393 is written in Scots.

95

first guardian and regent, there were still many nobles in the country who regarded the Bruces as usurpers and felt stronger allegiance to the Balliol line. Therefore, in July 1332, after Moray's death earlier in the year, when Edward, the son of John Balliol who had died in exile, sailed with a fleet into the Firth of Forth with the backing of Edward III of England and landed at Kinghorn in an effort to gain the throne, he had some initial success. It was, however, short-lived – he had himself crowned king in September but was out on his ear by the following December. In a second campaign to regain the lost throne in 1333, the Balliol faction inflicted serious reverses on the forces of David II, particularly a catastrophic defeat at Halidon Hill, and in 1334 the boy king, now ten years old, was shipped to France for his own safety.

What followed was an echo of events 40 years before. Sir Andrew Murray, the son of the Sir Andrew de Moray (the spelling of names was not fixed) who had fought with Wallace, was appointed guardian and regent by the supporters of the young king David. Born in 1298 after his father had died in the independence struggle, the younger Sir Andrew married a sister of Robert Bruce and acceded to the lordship of Bothwell in addition to his estates on the Moray Firth. Now he took on the leadership of the pro-Bruce faction as civil war again gripped the country. When Edward Balliol personally withdrew from Scotland he left his forces under the command of David de Strathbogie, earl of Athol, who, as he was a Comyn, presented another echo of the earlier conflict. Sir Andrew Murray and his army caught up with Athol's forces on Saint Andrew's Day 1335 and in a sharp engagement on Culblean Hill near Kildrummy in upper Strathdon gained the victory needed finally to secure the throne for David II. It did not bring an immediate end to the fighting but fairly soon afterwards the English monarchs, now started on what would prove a long war with France, lost interest in Edward Balliol and his cause. In the aftermath of Culblean, the widow of the earl of Athol took shelter in the fortress of Lochindorb and sent messengers to seek help from the English. A relief force duly arrived by sea, broke the siege that Sir Andrew Murray had thrown around the loch-bound stronghold, laid waste the country around Elgin and set off to blaze a trail of devastation around the Cairngorms and down through coastal Aberdeenshire. After that reverse, Sir Andrew Murray led the struggle until his death in 1338 in his castle at Avoch, after which the leadership of the loyalists fell to Robert the Steward, the son of Robert Bruce's daughter Marjory and her husband Walter the Steward. In

1341 David returned from his exile and launched a campaign of raids into northern England. In 1346, at Neville's Cross near Durham, he was seized by the English and kept in captivity for 11 years until a hefty ransom was paid by the Scots for his release.

The continual political unrest with which burgesses and peasants had to contend brought woes aplenty, but these were compounded by a long period of deterioration in the climate. During the fourteenth century conditions turned cooler and wetter across Europe, with a spurt in the growth of peat bogs and an increasing frequency of storms in the North Sea. In Norway and in northern England and Wales farming settlements retreated from higher ground to lower altitudes. We lack good evidence for this in Scotland, but it is likely that a similar response to worsening climate took place here. We can only make crude guesses as to the size of the population of the Moray Firth region in the Middle Ages. It may have passed the 1 million mark for the whole realm by the time of Bruce's victory at Bannockburn[6] and, accepting the proportion of a quarter of this total for the region, as described earlier, would give a population of over 250,000 around the Moray Firth. In 1315, the year after Bruce's victory at Bannockburn, harvests failed across the Continent, precipitating a famine that lasted in some places for three years. Growing conditions varied from year to year after that – the 1320s and the 1380s appear to have enjoyed warmer, drier summers – but the risk of dearth was never far away. Seven very severe winters occurred in the 1430s in England, and again there was famine. Generally through the fifteenth century severe winters and wildly varying harvests are recorded. In his classic study of historic climate, H.H. Lamb noted the paucity of data directly pertaining to Scotland but argued that the worsening climate contributed to the unsettled state of the country, as changes in the seasons would have exacerbated any competition for resources.[7]

Many rural people must have lived continually close to the breadline. Grain yields for Scotland were unlikely to have been greater than 1:3 or 1:4, that is, for each sown ear of oats or bere, the six-rowed variety of barley common to the north, the harvest would return three to four, and that in a good year.[8] Modern yields are over five times as much. We have to assume that the diets of the ordinary peasantry were based on only a few crops, with oats being of overwhelming importance, so much so that in the north to this day the crop is usually referred to in the country simply by the generic word 'corn'. There is almost no information about the rural diet for several centuries to come but what pertained in 1790 is highly likely to have been true for the Middle Ages. In describing the diet of his flock in 1794, the minister of King Edward parish in Buchan wrote: 'Animal food is rarely an article in the bill of fare but [except] on holidays . . . oats, bear and pease-meal, potatoes, and other vegetables, with milk, constitute the ordinary fare of the bulk of the people'.[9] Of course the potato was unknown in the Middle Ages but, that apart, the diet described was of long standing. The minister of the parish of Speymouth went into more detail: every day of the week, the peasants ate oatmeal porridge for breakfast, the oat dish called sowens at midday and greens or cabbage boiled with oatmeal for supper, and drank milk or beer – for the Middle Ages, substitute ale. On Sundays the menu comprised barley broth with some meat in the winter and butter in the summer; and fish was eaten by those who lived near the coast.[10] Add some cheese and bannocks made from oat- or bere-meal, and there we have a diet almost entirely dependent on a handful of crops and some dairy produce. Oats have a higher protein content than other grains typical of Europe and so long as enough is eaten can support decent health. The danger lay in the low yield. Farming in the Middle Ages provided little margin for catastrophe. The impact of any fall in production as the climate deteriorated would have been compounded disastrously by the outbreak of the Black Death in 1349, which may have wiped out more than a quarter of the people, although the north may have had some protection by its relative isolation. The plague never quite went away, adding the misery of regular epidemics to the evils attendant on harvest failure. By 1500 the national population may have fallen under 700,000, with about 175,000 in the Moray Firth region.

In northern latitudes prone to bad weather and harvest failure animals would have provided a more secure source of food than plants. It is no surprise to find the wealthy and powerful ensuring adequate access to animal protein, whether from domestic or wild sources. Fishing rights – who had them and what they were worth – feature hugely in charters of the period. All the large Moray Firth rivers were abundant in salmon, but sea fishing must have also been important, although documentation about it remains scanty until more modern times. Rents in the Middle Ages often specified numbers of poultry in the agreed list of payments in kind. There was also hunting, but the pursuit of game was by this time often restricted to the top of the feudal hierarchy. There is another social group who may also have been better equipped to cope with climate change and crop failure – those who practised pastoral agriculture. This meant largely the people who lived in the uplands, in the glens and straths away from the fertile margins of the Firth. As long as they had access to enough acres to graze their cattle and sheep, they could keep them alive for most of the year and store a summer's worth of butter and cheese as extra provender to see them through the lean months. Writing in 1380, the cleric John of Fordun noticed the distinction between lowland and upland: 'Scotia has tracts of lands bordering on the sea, pretty level and rich, with green meadows, and fertile and productive fields of corn and barley, and well adapted for growing beans, pease and all other produce . . . But in the upland districts and along the highlands the fields are less productive, except only in oats and barley. The country is there very hideous, interspersed with moors and marshy fields, muddy and dirty; it is, however, full of pasturage grass for cattle . . .'[11] Surplus cattle were traditionally slaughtered at the time of Martinmas, early November; those kept alive had to exist on meagre fare, mainly straw, unless the milder conditions found near the sea permitted survival outdoors where a foraging cow might find a bite. Writing of livestock on Skye at the end of the seventeenth century, Martin Martin has this to say: 'The common work-horses are exposed to the rigour of the season during the winter and spring; and though they have neither corn, hay, or but seldom straw, yet they undergo all the labour that other horses better treated are liable to. The cows are likewise exposed to the rigour of the coldest seasons, and become mere skeletons in the spring, many of them not being able to rise from the ground without help; but they recover as the season becomes more favourable and the grass grows up . . .'[12] Travelling in the Highlands in the early eighteenth century, Edmund Burt opined that spring was a bad season for the poor in the hills: '. . . for then their provision of oatmeal begins to fail and for a supply [of food] they bleed their cattle and boil the blood into cakes which, together with a little milk, and short allowance of oatmeal, is their food'.[13] In short, although life would have been perilously hard, the pastoral economy which had deep roots reaching into the Celtic past could have given the people who lived by it a significant margin of survivability in lean times. Although the emergence on to the historical stage at this time of many of the Highland clans familiar to us now may be partly an accident of record-keeping, it may also be the case that the old social and pastoral system of the clans could have been given a renewed vigour in the fourteenth century. Their way of life may even have appealed to the oppressed or fearful, and it may well have been wise for poor men to attach themselves and their families to the following of a lord or chief who could provide food and protection, one way or another, in return for loyal service.

The need for acreage for crops and for livestock meant that land was equivalent to wealth. Territory had always been important to the feudal lord and to the clan chief but, with a harder climatic regime and increased competition for land and what it could provide, the importance grew more intense. A circle of needs came into play – a lord needed territory, to keep territory he needed men, to keep men he had to provide for them, to provide for them he needed territory or rather the produce of territory, especially the produce that could be easily moved. In the Highlands this meant livestock. The old tradition of cattle-lifting, of raiding the neighbours to carry off their livestock, had deep roots in the Celtic past but it too may have been given a renewing surge by the conditions of the late fourteenth and fifteenth centuries. This may lie behind John of Fordun's often-quoted account of the differences between Highland and Lowland: 'The people of the coast are of domestic and civilised habits, trusty, patient and urbane, decent in their attire, affable and peaceful, devout in Divine worship . . . The highlanders and people of the islands, on the other hand, are a savage and untamed nation, rude and independent, given to rapine . . .' Fordun noted the Scots–Gaelic split and other obvious differences between the lowland and upland peoples, but in the midst of his apparent anti-Gael stance offers a more nuanced judgement. The Scots-speaking lowlanders, for all their compliant civility, are 'yet always prone to resist a wrong at the hand of their enemies', and the wild highlanders are, nevertheless, 'faithful to their king and country, and easily made to submit to law, if properly governed'.[14]

Taking his name probably from his birthplace in the Mearns, John of Fordun was a cleric in Aberdeen for at least part of his life. He would have been well informed about the social divisions in Moray and this region may have been in his thoughts when he put his views down on parchment. Unfortunately he has nothing to say about the state of affairs further north where a significant cultural shift may have been happening during his lifetime. It must have been during the fourteenth century that the Norse tongue spoken in Caithness began fatally to lose ground to a Caithness form of Scots and to Gaelic. Almost nothing is known about this process, although it was probably very similar to the later, better understood displacement of Norse by Scots in Orkney and Shetland. It has been suggested that Caithness may for a time have been trilingual with

Norn, Scots and Gaelic all present,[15] but the scenario of Norse giving way to Gaelic in the west of the county and to Scots in the north-east, with the implication that Gaelic was never the common tongue in the north-east, has had wider acceptance. The distribution of place names in Caithness with a marked frontier zone running across the county between Gaelic- and Norse-derived names supports this more prevalent view.

The origins of some of the clans and kindreds that materialise from obscurity around the Moray Firth are unknown. We can speculate that some may have had deep roots in the area whereas we know that others were recent incomers or representatives of a family that had in a few generations been able to enlarge its network of possessions. Dr W.F. Skene, one of the great Victorian scholars of the early history of Scotland, thought that the Mackintoshes were descended from the original clan Chattan and ultimately from one of the Gaelic-speaking kindreds. This is an attractive notion but it cannot be confirmed. Another tradition, recorded in a Mackintosh manuscript of 1679, has the Mackintoshes springing from Shaw, a son of the third earl of Fife who had fought for Malcolm IV in his defeat of the men of Moray in 1163 and been rewarded with the constable-ship of Inverness Castle, the lands of Petty and Breachley and the forest of Strathdearn. This is highly plausible but also, sadly, difficult to confirm. The oldest surviving charter among the Mackintosh muniments is dated 5 October 1442 and records the granting to Malcolm Mackintosh of the lands of Meikle Geddes and half of Rait, including the castle. The name Mackintosh means son of the chief – *toiseach* – or thane, as Shaw, son of the earl of Fife, could indeed have been. The derivation of the name Chattan, now always assigned to a federation of clans among whom the Mackintoshes were the most powerful, has never been adequately explained.[16] The origin of the Munros in Easter Ross is likewise surrounded by tradition. One version of their history is that they descend from Irish followers of the earl of Ross and derive their name from Roe Water in County Derry but an alternative view is that they moved from Moray to Ross, possibly to escape from David I and his successors.[17] The oldest authentic record – of George Munro, identified as the fifth baron of Foulis – dates to around 1232, when his name appears as a witness to a charter by the earl of Sutherland to the archdeacon of Moray. The Mackenzies emerge from a kindred granted lands in Kintail by the earls of Ross, one of whom in the time of Robert Bruce had the name Coinnich and whose descendants bore the patronymic MacChoinnich.[18]

The Fearann Domhnuill stone in the grounds of the Storehouse at Foulis on the north shore of the Cromarty Firth commemorates one tradition surrounding the origin of the Munro clan – that it descends from Donald O'Caan, a prince of Fermanagh, who came to fight the Norse alongside Malcolm II in the early eleventh century and was rewarded with the territory between the Alness river and Dingwall, the land of Donald – *Fearann Domhnuill*.

During the fourteenth and fifteenth centuries, several families from the south moved north to occupy prominent positions, filling the power vacuum left by the demise of the Comyns, finding their place among the clans and the kindreds who had come ahead of them. Among them were the Chisholms, of Norman-French origin, who appear first in the Borders in around 1254 and moved to their lands in Strathglass a century later.[19] William Sinclair, a scion of the Norman-French family that had first been granted lands in Midlothian in the time of David I, acquired the earldom of Caithness in 1455. We find Brodies for the first time in Moray in 1311 as allies of Robert Bruce. Ogilvies became the lairds of Deskford in the 1430s and a powerful kindred in Banffshire. Keiths from the family of the hereditary earls marischal of Scotland moved up the east coast from the Lothians via Dunottar to Caithness, where by marriage to a daughter of the Cheynes they acquired the lands of Ackergill.[20] An Oliphant similarly leapt the Firth to acquire through marriage the Caithness properties of Berriedale and Oldwick.

By tradition the rallying place for the clan of Sutherland on the old bridge at the north end of Golspie. The inscription reads *Morfhear Chatt de Cheann na Droichaite bi Gairm Chlann Chattach nam Buadh* – roughly 'the great man of Sutherland from the head of the bridge calls the clan to battle'. Note how the Gaelic preserves the ancient Pictish name of Cat, also adopted by the Norse when they named Caithness as *Katanes*. The bridge was built in 1808, long after the traditional days of the clans.

The Freskin clan continued to strengthen their grip around the Firth. William, the fifth earl of Sutherland, married Margaret, the sister of David II. Their son John, born in 1346, became third in line to the throne after Robert the Steward and, if he had not died in 1361 at the age of only 14, and had not the Steward ensured all his own offspring were legitimated, the house of Sutherland may have become the next royal dynasty. In 1350 Robert the Steward's daughter Margaret married John Lord of the Isles and, in 1355, Robert himself married the sister of the earl of Ross, Euphemia, the countess of Moray and the widow of Earl John Randolph. After the death of David II in 1371, Robert the Steward finally became king, aged 54 and bearing the unpromising nickname of 'Auld Blearie'. Nevertheless, he reigned for nearly another 20 years, dying in 1390 to be succeeded by his son, John, who styled himself Robert III. Beset by physical and psychological ailments, the two Roberts failed to keep a firm grip on their subjects; the country slid into lawlessness and suffered some humiliating defeats by England. As we have seen, however, some of the nation's problems may have been due to environmental factors beyond royal control.

In the latter decades of the fourteenth century the politics along the shores of the Moray Firth from Ross to the Deveron were dominated by three men. At the head of the church was Alexander Bur, bishop of Moray. His name suggests he belonged to Aberdeenshire, and he became bishop

The ruined walls of Lochindorb, a residence of the Wolf of Badenoch, crouch on their island sanctuary amid an expanse of moorland.

in 1362 during the reign of David II. His secular counterpart was Thomas Dunbar, the sixth man to hold the earldom of Moray after it had been forfeited on the fall of the Comyns and Robert Bruce had awarded the title and the estates to his faithful Thomas Randolph in 1312. Two Randolph sons had followed in due succession but, in the fourth generation, the inheritance had come to a daughter, Isabella, who had married Patrick, earl of Dunbar. The Dunbars were senior nobility, able to trace their forebears back to a Northumberland earl who had fled to Scotland to escape William the Conqueror and who had been granted the Lothian lands with the castle that gave the kindred its surname. Patrick's son John, who married Marjory, another daughter of the future Robert II, was granted the earldom of Moray in 1372, and his son Thomas inherited it in February 1392. The earldom had been reduced in size by the removal of Lochaber, Badenoch and Urquhart Castle but it was still an extensive and fruitful province.

The third man in this triangle of power is the one who is best remembered by the nickname later given him in acknowledgement of his misdeeds – the Wolf of Badenoch. In his own time, he was known as the seneschal or steward. He was Alexander Stewart, an illegitimate son of Robert II, and was probably granted the lordship of Badenoch by his father where he first appears at his base at Ruthven in August 1370 when he signed letters promising his protection to all the men and property of the bishop of Moray in Strathspey and Badenoch, a pledge that seems to have bothered him little in the years that followed. A royal grant in March 1371 conveyed Lochindorb to him, and he later extended his possessions by leasing Glen Urquhart from his brother and obtaining the barony of Stratha'an, the valley of the Avon on the north-east side of the Cairngorms. These are all upland territories, better suited to pastoral than arable agriculture, but major and more fertile prizes soon accrued to Alexander. In June 1382 he married Euphemia, the widow of Sir Walter Leslie, earl of Ross. As well as the fertile acres around the Cromarty Firth, the earldom included the barony and sheriffdom of Nairn and the barony of Kineddar in Aberdeenshire. The latter addition to his portfolio of properties made Alexander also earl of Buchan. Both the Dunbars and Alexander Bur must have watched this encirclement and the swelling power it represented with growing unease. To make matters worse, Robert II had appointed Alexander the seneschal, already the sheriff of Inverness, as royal lieutenant for all the country on the north side of the Firth and for that part of Inverness-shire that lay outside the earldom of Moray. In his study of the seneschal, Alexander Grant suggests that it is unsurprising to find the Dunbars spending most of their time in the south – to escape a deteriorating political as well as a natural climate.[21] This left Alexander Bur the single figure in the north with the clout to confront the seneschal, although Bur may also have nursed ambitions of his own to regain the authority the bishopric had enjoyed a century before. The bishop and the seneschal fell out over the superiority of church lands in Badenoch. In February 1386 they met in the house of an Inverness burgess, Thomas, son of John, and reached a settlement. Three years later, Euphemia of Ross fled from her husband in a cloud of complaints about ill treatment – the seneschal had deserted her to live with a mistress, with whom he had several children – and the bishop excommunicated him. In the summer of 1390 the seneschal gathered his followers and descended on the lowlands of Moray to wreak vengeance by sacking and burning the town of Elgin and the great cathedral that Walter Bower described in his history as an ornament to the whole country. The reek from this infamous crime has obscured the fact that the seneschal also attacked and burned Forres that same miserable season. He was eventually forced to appear in penance for this monstrous attack on the church and receive absolution. The contemporary chronicler Andrew of Wyntoun recorded the sack of Elgin in verse:

Today the gaunt walls of Elgin Cathedral are only a reminder of its magnificence. Although there was much rebuilding after it was sacked and burnt by the Wolf of Badenoch in 1390, the cathedral was abandoned after the Reformation in 1560 when the focus for worship in the burgh shifted to the parish church of St Giles and soon deteriorated through neglect and the removal of building materials. In 1637 a gale removed the roof of the choir and in 1711 the central tower collapsed. Finally in the early 1800s the importance of the cathedral was realised. Today it is in the care of Historic Scotland.

That ilk yhere eftyr syne
Brynt the kyrk wes of Elgyne
Be wyld wykkyd Heland-men
As wedand [frenzied] in thair wodnes [madness] then . . .[22]

If proof were needed, here was ample evidence for the nature of 'wyld wykkyd Heland-men'. This division between Highland and Lowland was characterised probably less by language than by perceived differences between urban and rural, coastal and upland people, tillers and pastoralists, a regard for authority and lawlessness, even civilisation and unreconstructed barbarism. This division and the associated denigration of the Highlander was to colour the Moray Firth region and much of the rest of Scotland during the coming centuries.

Of our protagonists, Bur died at Spynie in 1397. The seneschal, having lost much of the land he had earlier controlled, survived until 1405. His tomb is in Dunkeld Cathedral. Euphemia of Ross ended her days as abbess of Elcho where she passed away in 1398. She was buried in Fortrose Cathedral. Her son, Sir Alexander Leslie, became the earl of Ross but he died in 1402 and the earldom passed to his younger sister, Mariota, who married Donald of Islay, Lord of the Isles. During the fourteenth century the Clan Donald, descended from a Norse-Celtic kindred, grew in power and influence in the western Highlands and the chief began to style himself *Dominus insularum*, Lord of the Isles. They had already made their presence felt around the Moray Firth. One of the family, Alasdair Carrach, held the lordship of Lochaber. In 1394 at Cawdor, Thomas Dunbar, earl of Moray, seeing the Donalds as allies against the seneschal, reached an agreement with Carrach over the protection of church lands in Moray. In 1398, Carrach, possibly feeling he was not gaining much from this agreement, decided to extend his control up the Great Glen, seized Urquhart Castle and partitioned the extensive holdings of the bishop at Kinmylies to the west of Inverness between his own allies. Some land was given to John de Cheshelm of the Aird, and associated Ness fishings were granted to an Inverness burgess, John Qwhyte, an occurrence that suggests how alliances of mutual advantage could bridge the gap between urban and rural societies. The bishop made use of the church pulpit to warn de Cheshelm off with threat of excommunication. The dispute between the bishop and the clan chief rumbled on unresolved until, in July 1402, Carrach himself sacked Elgin.

Through his marriage to Mariota, Donald claimed the earldom of Ross but was denied it by the crown. Finally, in 1411, after failing to persuade Lord Lovat, the chief of the Frasers, to join him, Donald led his men east to realise his ambitions by force. According to the Wardlaw Manuscript, written over 200 years later but recording long-standing oral tradition, Donald 'marches through, pillages and plunders all before him, attacks Inverness, burns the bridge, the famousest and finest off oak in Brittain, burns most of the town, because they would not rise and concurr with him'.[23] The men of Clan Donald, 1,000 or so in number, carried on through Moray in the direction of Aberdeen. On the braes of Harlaw to the north of Inverurie the clansmen were confronted by a force led by the earl of Mar, fought an indecisive skirmish and turned back for the west. Mar followed them to ensure they kept on the move. In 1412, he received £100 from the royal treasury for the repair of the castle at Inverness, along with £32 10s 3d for lime, supplies and transport of timber.

The rampages of the Wolf of Badenoch and the Clan Donald are two major examples of the political warfare common in the north in this period. There were also many more local feuds. In 1428, in an effort to impose a degree of order on the Highlands, James I resorted to a ploy to bring the chiefs and magnates to heel. Perhaps the king felt he had no choice, seeing neither rhetoric nor straight military confrontation as feasible options, but his plan – summoning all the leading men to a parliament in Inverness and, as they arrived one by one, arresting and confining them – seems remarkably ill judged. The chiefs never forgave him for the trick. Among the captives were Alexander, Lord of the Isles, the son of the man who had led his followers to Harlaw, Angus Dubh Mackay and his four sons from Strathnaver, and Kenneth Mor Mackenzie. Three of the regional warlords were executed for crimes and the others were constrained in various ways to respect the king's peace. One of Angus Dubh's sons was confined as a hostage for seven years on the Bass Rock to ensure his father's good behaviour. Alexander of the Isles refused to be won over by the king's arguments and after his release gathered his followers to sack Inverness, the scene of his humiliation, in 1429. Relations with the crown improved a little for the Lord of the Isles after James II succeeded his father on the throne. Alexander was granted the earldom of Ross and in 1438 was appointed as sheriff in the Highlands, a post he held until his death in Dingwall in 1449. Alexander was succeeded as Lord of the Isles and earl of Ross by his son, John. Married to Elizabeth Livingstone, the daughter of the earl of Crawford, perhaps for political reasons, John was drawn into rebellion when Crawford fell out of favour at court. In March 1451 the Clan Donald under John seized the castles of Urquhart, Ruthven in Badenoch and Inverness. Crawford was defeated by the earl of Huntly at the battle of Brechin in 1452, and the castles at Inverness and Urquhart – Ruthven had been demolished – were annexed to the crown in 1455. John of the Isles escaped any major sanction but the Gordons of Huntly became the hereditary keepers of Inverness Castle.

The origins of the Gordons are unknown, although there is a strong possibility that the first of the kindred was a Norman knight rewarded with land by Malcolm Canmore in the parish of Gordon in Berwickshire. The kindred was content to stay in the Borders until in 1319 Sir Adam Gordon was granted the forfeited Comyn territory of Strathbogie by Robert Bruce. The main stem of the family moved north and became in time a significant political force. The great-grandson of the first Sir Adam fell at the Battle of Homildon Hill in 1402, leaving an only daughter as his heir. In 1408, she married Sir Alexander Seton – their children assumed the name of Gordon. Sir Alexander had ambitions and, according to tradition, built up his following by offering gifts of oatmeal to all who would take the name of Gordon, giving rise to the disparaging nickname of 'bow o' meal Gordons' for some branches of the clan. This story may have more to do with a widespread resentment of burgeoning Gordon power but it could also reflect economic and ecological reality at the time. The real wellspring of the Gordons' rise lay in their seizing of opportunities through keeping close to the monarch, marrying well and being appointed to important positions in state and church. The Setons' son, Alexander, when he chose a daughter of the chancellor of Scotland to be his second wife, was created the first earl of Huntly some time before July 1445. He chose as his territorial title what was then a small village and received grants of the lordship of Badenoch and other estates in Inverness-shire and Moray.

The second earl, George Gordon, acquired more lands, was appointed the king's lieutenant with judicial authority over all between the Forth and Orkney in 1476 and served as chancellor from 1498 to 1500. He divorced his first wife, a Dunbar from the family of the earls of Moray, and for his second married a daughter of James I. That union also ended in divorce but did see the birth of the third earl of Huntly, Alexander Gordon, who continued the family custom of extending the acreage under their name – in his case, acquiring Strathavon in Banffshire, and the Braes of Lochaber. Alexander commanded one wing of Scottish troops at Flodden but survived the carnage to die peacefully in 1524. His son, the fourth earl, George Gordon, was born in 1513. Alexander Gordon's younger brother, Adam, married Elizabeth, countess of Sutherland, in 1500. First Elizabeth's father and then her brother, the rightful heir, were found by inquests under Huntly's control to be insane and unfit to hold an earldom. Perhaps the father was senile before he died in 1508, and perhaps the son was a simpleton – he died in 1514 – but suspicion still attaches itself to the way they were sidelined. Elizabeth's second brother, Alexander, proved to be tougher to handle but he was killed. The outcome, in any case, was that Adam Gordon assumed the earldom of Sutherland by means many still regard as malicious trickery, giving the Gordons firm footholds on both sides of the Moray Firth.

The acquisition of Sutherland brought the Gordons up against the territory of the Sinclairs in Caithness. This family had been in the far north for several generations, first in Orkney and then in 1455 being granted the earldom of Caithness. The second earl lost his life at Flodden in 1513. James V appointed the Sinclair earls as hereditary sheriffs and justiciars of the two northern counties in 1527, in an effort to impose some order on territories riven by feuding and cattle-raiding, activities in which such clans as the Mackays displayed notable zeal. After the third earl fell in a bloody police action in Orkney in 1529, the offices of sheriff and justiciar were eventually invested in his son, George Sinclair, the fourth earl, in 1567. This placed the earl of Sutherland in theory under the jurisdiction of the earl of Caithness. Looking back at the records, it is at times difficult to distinguish between the pillaging associated with a quarrel or a cattle raid and the activity of George Sinclair in his role as justiciar. His official position did not prevent him from becoming involved in a long-running feud with the house of Sutherland that can only be outlined here. This clash had its roots in a curious incident in June 1567 – when John Gordon, the earl of Sutherland, and his wife were poisoned as they ate dinner in their castle at Helmsdale. A certain Isabel Sinclair, who attended the earl and his wife on the fateful day, was found guilty of the crime, but many Gordons felt that the mastermind behind the deed was really George Sinclair. There is no evidence that the Caithness earl was involved but, once the Sutherlands were in their graves, he wasted little time before acquiring the wardship of Alexander Gordon, the dead earl's son, and marrying him off to his daughter. In 1569, with the assistance of some Murrays in Dornoch, Alexander Gordon escaped across the Moray Firth to Huntly territory. John Sinclair, the son of the earl of Caithness, now formed an alliance with the Mackays and attacked Dornoch to teach the Murrays a lesson, sacking the small town in 1570. The cathedral was left in a ruined state until well into the following century. Alexander Gordon later obtained a divorce from George Sinclair's daughter and married anew, this time to a daughter of the earl of Huntly. George Sinclair died in 1582 to be succeeded by his grandson, and this presented the Gordons and others with an opportunity to attempt a curtailment of Sinclair power through the Privy Council. The effort was successful but it provoked a fresh round of lawbreaking in Caithness as Sinclair followers sought vengeance. In 1585 the Privy Council appointed the earl of Huntly to mediate between the Sinclairs and the Sutherland Gordons, negotiations that concluded, among other things, that the Gunns, a small clan living mainly in the Kildonan area, had been a main cause of trouble between the two houses, a curious conclusion that was probably related to access to lands on the border between Caithness and Sutherland. The two earls now combined forces to suppress the Gunns but the alliance did not last long and in 1587 the Sinclair–Sutherland feud broke out afresh. Long-distance raids were made into each others' territories until, in the spring of 1589, the Sutherland men, eagerly assisted by Mackays and Macleods from Assynt, swept along the east coast to capture and pillage Wick itself. This great raid became known in Gaelic as *latha na creach mor,* the day of the great spoil. The earl of Caithness sat it out in his impregnable stronghold at Girnigoe and then appealed to James VI to make Wick a royal burgh. The king had already renewed the charter for Tain in 1587 and the appeal in relation to Wick, although avowedly to develop trade in the north, was raised to gain for the battered town a degree of royal protection. James opportunely happened to tour that summer from Aberdeen to Cromarty and no doubt it was during this progress that the Sinclair request was made. James signed the charter in September. This did not bring an immediate end to the Sinclair–Sutherland feud but after a few more tit-for-tat raids it gradually dwindled into a relatively harmless rivalry.

# CHAPTER 9
# STRUGGLES IN KIRK AND STATE

'Betwix Ros and Murray the land crukis in with ane gret discens and vale, in quhilk fallis five rivers, Nes, Nardyn, Findorn, Los and Spay.' With these words in a book published in 1527, Hector Boece tried to convey the shape of the inner Moray Firth.[1] Boece was, like so many academics and scholars in his time, at home almost anywhere in Europe. He was professor of philosophy in the University of Paris in 1497 when he received an inducement he could not resist – to return to Scotland to help establish a new university in Aberdeen. Here he wrote his great work *Historia Gentis Scotorum*, the history of the Scottish people, which appeared first in Latin but was soon translated into French, English and Scots. Another international Scot and a contemporary of Boece, John Mair or Major had published another history of Scotland only six years before. Boece has more to say than Mair about the Moray Firth but both men are patriotic and mingle fact with fabulous story. Although their works were products of the Renaissance sweeping Europe, the shadow of the Middle Ages still lay across them. Boece noted how Sutherland was profitable country for livestock and grain, how the name Moray once referred to 'all the boundis betwixt Spay and Nes to the Ireland seis' but was now reduced to the country between Spey and Kessock, how in Moray itself there was 'gret aboundance and fouth of quheit, beir, aitis, and siclike cornis, with gret plente of nutis and appillis, . . . gret fouth of fische and speciallie salmond', how the loch of Spynie had grown so shallow and weed-choked that most of the fish had gone, and how there were plenty of sheep and cattle in Banff and Buchan. He also accepted that herring had vanished from the mouth of the Ness for some offence made against a saint, that the bones of Little John could be seen in the kirk at Petty, and that rats could not survive in Buchan. He noted how the blessed bones of Saint Duthac were greatly venerated in Tain, now a centre of pilgrimage, famously attracting James IV to pray before the sacred remains every year from 1493 until his last visit in 1513, one month before the disastrous battle at Flodden.

The books by Mair and Boece set the scene for us for what turned out to be a 250-year-long struggle for the soul of the nation. Much of the history of the Middle Ages around the Firth remains obscure but from what we can tell, in the early 1500s the traditional, long-standing contests between magnates and monarchs over who held sway in the various regions now became complicated by the clash of ideas, first in the church about the interpretation of scripture and then in society in general about how a nation should be governed. First came the Reformation, then in the turbulent seventeenth century the wars of the Covenant and the Wars of the Three Kingdoms, and finally in the eighteenth century the upheavals of the Jacobite risings. By the end of this period society had changed in significant and major ways. And it all took place against a backdrop of severe fluctuations in climate and harvests. The changes in the governance of the church were fitful – episcopacy was abolished under the Covenanters only to be restored with Charles II and abolished again in the Church of Scotland in 1690. The line of Stewart/Stuart monarchs lost the throne after much struggle and was replaced by a constitutional monarchy and a royal family imported from Hanover, seen as German although it had a few Stewart genes in its bloodstream. Among the nobility and the clan chiefs and lairds who emulated them the sword eventually gave way to the legal writ as a means of settling disputes. The blood feuds and cattle-lifting that bedevilled rural life for the peasantry died out after first being afforded some dubious legitimacy when one clan was licensed by a helpless central government to bring chastisement by fire and sword on another. The burghs maintained themselves as best they could as pockets of mercantile society, with a role out of all proportion to their size. To some extent they were neutral territory, places apart from the surrounding struggles between kindreds and clans where rival leaders could meet to thrash out their differences. The burghs were portals for trade and traffic with the outside world but with regular markets, stocks of commodities and portable wealth they also became natural targets. Time, however, was on their side, for they also cradled a new element in society – the emerging class of ministers, lawyers, merchants, and eventually doctors, teachers, bankers and other professionals – that would acquire increasing influence and importance.

The pressures within the church for reform of doctrine and practice reached a crisis point in 1517 when the German priest Martin Luther

An ancient tradition that still lives is manifested at the Clootie Well near Munlochy on the Black Isle. This spring is endowed with healing properties and is steadily frequented by people who tap its power by dipping a rag or cloot into the water, and hanging it up to resolve a problem or answer a wish as it rots. This belief in springs and wells having magical powers extends from the pre-Christian Celtic period.

nailed to the door of the church in Wittenberg his protest against the sale of indulgences. The tenets of Lutheranism spread quickly across northern Europe. In 1525 Parliament in Edinburgh passed an Act to ban the import of pamphlets on what it called 'the dampnable opunyeounis of heresy spread . . . be the heretic Luther' that merchants and sailors were bringing to the shores of Scotland.[2] Attempts to reform the church from within failed to stem the momentum for change, and such drastic measures as the burning of Protestants at the stake as heretics created martyrs for the cause. A climax was reached on 1 August 1560 when Parliament officially declared Scotland to be a Protestant nation. In its final stages, the Reformation seems to have followed a fairly peaceable course in Scotland. The outbursts of violence, such as the looting of two priories in Perth, were enough, however, to persuade churchmen to look to their own safety. The records are patchy for events around the Moray Firth where the leaders of the various religious establishments took a range of courses. The Dominican friary in Inverness was not attacked although the prior, clearly fearing it might be, handed over valuables to the safe keeping of the provost and bailies, in whose hands they seem later to have disappeared. John Fulford, the prior of the Carmelite community in Banff, took pre-emptive action in 1559 by granting to George Ogilvie of Castleton a tack of 'all and haill our place beside Banff, with yaird, orchard and other townis contenit within the stain wallis' for 11 years for a rent of £6 every Whitsun and Martinmas to be paid 'to the prior or his successors in quhat strait that ewer thai be for the tyme be ressoun of this present contrawersie'. A warning of what could happen had come when someone had started a fire in the Carmelite premises, causing damage and 'manifest spuilzie' inside the church. Carmelite convents elsewhere in Scotland were also attacked at this time. In 1561 Fulford granted further lands to the Ogilvies, and in 1574 the crown annexed former Carmelite lands for King's College in Aberdeen.[3] Nicholas Ross, the lay commendator or administrator of the abbey at Fearn and also the provost of Tain, obtained a bond from Alexander Ross of Balnagown who pledged 'to mantene and defend' him.[4] Nicholas attended the crucial Parliament in 1560 that legalised the Reformation but before he set off for the south he had Alexander Ross promise 'upon ye relequies of tane' to keep the said holy relics within his own house. The relics were described as 'ane hede of silver callit sanct Duthois hede his chast blede [breast bone] in gold and his ferthyr [portable shrine] in silver gylt wt gold', valuable items in monetary terms, priceless in religious significance. Nicholas, who must have been in favour of reform while still holding respect for older practices, and Robert Munro of Foulis, the representative in Parliament of the barons of Ross, voted for the adoption of Protestantism. Walter Reid, the abbot of Kinloss, joined the Reformers and married. The abbot of Pluscarden, Robert Dunbar, hastily made provision from the abbey lands for his illegitimate children before his expulsion. Patrick Hepburn, the bishop of Moray, however, shut himself off in his castle at Spynie and defied the Reformers 'alike as to his estate and his morals', according to a later historian, until his death in 1573.[5] The bishop of Caithness and Sutherland, Robert Stewart, missed the crucial moment of the Reformation. He had been living in exile since 1545, after he had become embroiled in a feud between his brother and the supporters of the earl of Arran, but when he returned to Scotland in 1563 he resumed his clerical career by accepting the new dispensation. He kept his bishopric and was charged with planting kirks within his diocese, a task he carried out with diligence, earning the thanks of the General Assembly.[6]

The reactions to the Reformation among the common people must have been mixed. For some it would have been an unsettling time marked by the overthrow of cherished beliefs, for others a long-sought cleansing of the church. To the nobles, the burgesses and the lairds – to anyone in a position of some authority – it presented opportunities of enrichment, to grab or hold on to a share of the church's wealth. In 1562 three canons of Fearn complained to Nicholas Ross that the laird of Balnagown, Alexander Ross, had forced them to subscribe a charter supposedly drawn up by a former abbot concerning lands set in feu ferme to him. Nicholas went to law to fight Alexander's obvious attempt to acquire some land. He probably served as Tain's first Protestant minister before he resigned from his post in 1567 and died two years later. In his last years he had obtained letters of legitimation for his illegitimate offspring, including at least three sons for whom he bought the farm of Geanies from Balnagown.

The General Assembly recognised the need to provide for the clergy of the old church and decided in January 1562 that all the holders of benefices would continue to enjoy their revenues, although one-third of the income was creamed off to the crown, in effect a tax both to boost the state finances and to pay the stipends of the new reformed ministers. This measure was designed to satisfy everyone who had a stake in the extensive resources of the church, particularly the many nobles and lairds who had an interest in hereditary church offices or an eye or already a foot on church land. Collectors were appointed to gather revenue according to the rentals that all beneficed men were called on to produce. It took some time to appoint reformed clergy to all the parishes. In some cases the incumbent priest accepted the Reformation and stayed in place, in others lay readers or exhorters filled the posts until properly ordained ministers came along. On the Black Isle, for example, Alex Pedder, the vicar in Avoch, stayed on but John Anderson, the vicar in neighbouring Cromarty, refused to conform and the parish was served by a reader called James Burnet from 1569. Dornoch acquired an exhorter in Gaelic in 1567. Some of the new ministers may have been over-zealous – David Ray, appointed to Forres in 1563, was admonished before the General Assembly for not observing a properly decent order in his preaching and apparently leavening his sermons with invective.[7] In Inverness, in June 1561 a certain Johne Morison was found guilty of slandering David Rag, the town's first Protestant minister, by calling him a 'commone pulpit-flitter' and accusing him of molesting men's wives. Morison was duly punished by the burgh court but Rag was soon accused again by another man of adultery and threatening behaviour. Rag, who may have been a mendicant friar before the Reformation and a kind of

The view from Califer Hill across Forres towards the mouth of the Cromarty Firth, guarded by the Sutors of Cromarty. (*Jim Henderson.*)

Duncansby Head and its lighthouse define the northern corner of the Moray Firth area. At this point the North Sea joins the Atlantic through the narrow channel of the Pentland Firth, seen here extending west into the distance. (*Courtesy Moray Firth Partnership/by Scotavia Images.*)

A succession of rocky headlands extends west from Knock Head, near Whitehills. (*Courtesy Moray Firth Partnership/by Scotavia Images.*)

Aerial view of Banff and the harbour. (Courtesy Moray Firth Partnership/by Scotavia Images.)

The sandbars and shoals of the Gizzen Briggs guard the entrance to the Dornoch Firth. (*Courtesy Moray Firth Partnership/by Scotavia Images.*)

The mouth of the Spey. Garmouth and Kingston on the west side were the centres of the shipbuilding industry founded on pine floated down from the forests of upper Strathspey, whereas Tugnet on the eastern shore became a centre for salmon fishing. Here the six-chambered ice-house, now owned by the Whale and Dolphin Conservation Society, is open to the public. (*Courtesy Moray Firth Partnership/by Scotavia Images.*)

The lighthouse at Kinnaird Head in the lower part of the picture was built in 1787, one of the first quartet of such facilities erected by the newly created Commissioners of Northern Lights. It is now home to the Museum of Scottish Lighthouses. Just to the south of it lie the harbour complex of Fraserburgh, founded in the sixteenth century and the coastline of Aberdeenshire trending to the south. (*Courtesy Moray Firth Partnership/by Scotavia Images.*)

Sir John Sinclair of Ulbster, as portrayed by Sir Henry Raeburn in 1794–95. (*Scottish National Gallery.*)

George Dempster of Skibo, by John Opie.
(*Scottish National Portrait Gallery.*)

The deep water of the Cromarty Firth and the sheltering, narrow entrance of the Sutors made Invergordon an ideal base for the Royal Navy during the two World Wars. The same features today enable the town to service oil production platforms from the North Sea and host ocean-going cruise liners, welcome economic activity since the short-lived industrial phase of the 1970s. On the near side of the North Sutor the Nigg construction yard can be seen, matched by the town of Cromarty on the South Sutor. (*Courtesy Moray Firth Partnership/by Scotavia Images.*)

religious opportunist, disappears from the record and is replaced by a new minister by June 1565.

Some nobles, such as the earl of Caithness, stayed loyal to the old dispensation, no doubt to the annoyance of a General Assembly without the power to do much to affect them. The most notable adherents to the Roman church, though, were the Gordons of Huntly, whose power and leadership over an extended kindred and tenantry ensured that the north-east remained conservative in religious terms. George Gordon, the fourth earl, was a loyal supporter of James V and served as regent while James went off to France in 1536 to find a wife. After James's death in 1542, George continued to support his widow, Mary of Guise, and it was on his suggestion that the young princess Mary married the Dauphin. George successfully consolidated his position through bonds of manrent, marriage agreements and the outsmarting of rivals. James Stewart, the earl of Moray, died without an heir in 1544 and five years later the earldom was granted to Huntly by a grateful monarch. The Gordon ascendancy may have appeared unstoppable. The earl was nicknamed the Cock o' the North. It is ironic that it was to be another lineage of Stewart earls who would soon present the principal challenge to the Huntly position, a contest given an extra edge during the Reformation years by the Stewart adherence to the Protestant cause. The new James Stewart in our story was an illegitimate son of James V and, therefore, a half-brother of Mary Queen of Scots, a man on whom Mary came to rely heavily on her return to Scotland in August 1561. Unsurprisingly, given the religious and political tensions between them, Stewart and Huntly quarrelled. The ill-feeling was made worse when Mary took the earldom of Moray from Huntly to award it to her half-brother in January 1562. In August of that year Mary and James Stewart set off on a tour to the northern districts. In Aberdeen the young queen lodged with Huntly but became suspicious of his loyalty and declined to visit his castle at Strathbogie. When she and her retinue reached Inverness, the Gordon commander of the castle garrison, one of Huntly's sons, refused to allow the royal party to enter on the grounds he had received no instructions from his father. Another Huntly son was at that time at large in the Highlands with a price on his head and an armed following. To prevent any attempt to kidnap the queen, a large number of Fraser, Munro and Mackintosh clansmen, none of them lovers of Gordons and needing no second asking, hurried into Inverness to strengthen the royal party, finally escorting Mary as far as the Spey to see her safely on her way south again, before besieging and capturing Inverness Castle and hanging the misguided Gordon captain. Huntly was now in deep trouble and his protestations of loyalty did not save him from defeat in a clash of arms at Corrichie outside Aberdeen on 28 October. He died of a stroke either in the heat of battle or shortly after being taken prisoner, and then his embalmed corpse was carried to Edinburgh to be the main feature in a grisly proceedings in the Parliament in May 1563 when an Act of attainder and forfeiture was passed on it. Huntly's heirs and successors were barred from any office of rank or honour. Huntly's remains were eventually laid to rest in Elgin Cathedral, and his son and heir, also called George, was imprisoned to await a death sentence.

That may have been the end of Gordon power. The queen, though, saw fit to keep the young George alive, thinking of him as a possible useful counter to any untoward ambitions of her half-brother, now the earl of Moray. When the latter and his Protestant allies opposed Mary's marriage to Lord Darnley, George Gordon's moment came. He found his prison door unlocked, his title as Lord Gordon restored and, when he brought out his followers from Strathbogie in support of the queen against Moray, received in 1566 the earldom of Huntly as well. Moray and Huntly both became deeply involved on opposite sides in the unravelling of the reign of Mary Queen of Scots, the feud between them playing out on the national stage. Mary appointed Huntly her lieutenant in the north in September 1568. Moray, as regent for the young James VI, tried to bring Huntly to heel, even levying a fine on the Gordon earl. In January 1570 Moray was assassinated in Linlithgow by another supporter of Mary – he has the dubious distinction of possibly being the first political figure to be murdered by a firearm – opening the way to Huntly to reassert his authority in the north-east and in Moray.

From his regional capital in Aberdeen, Huntly governed the whole swathe of country as far as Inverness, funding this exercise of power with court and church revenues and in effect operating his own foreign policy – he gave shelter to Catholics and refugees from a failed rising in England against Elizabeth – in defiance of Edinburgh. With kin holding the earldom of Sutherland, Huntly was strengthened by access to manpower and a refuge on the northern side of the Firth. He spent most of his time and energy on national politics until it became clear that Queen Mary, who had fled to England in 1568 to become a prisoner of her cousin, was not coming home again. In February 1573, Huntly and his fellow pro-Mary supporters signed the pacification of Perth, accepting the rule of James VI and the reformed kirk. The old warrior finally succumbed to a stroke apparently brought on by a strenuous game of football in Strathbogie in October 1576.

The Moray–Huntly feud continued into the next generation. The fifth Gordon earl was succeeded by his son, also called George. Rivalling him in the north was James Stewart, the son of James, the former queen's half-brother, and now the second earl of Moray in its latest incarnation. Late in 1590 Moray took the side of the Grants of Ballindalloch and the Mackintoshes in a feud with Huntly. James VI tried to force a settlement on the warring magnates, ordering Huntly to stay east of the Spey and Moray to the west. Moray chose to link himself with his wife's cousin, Francis Stewart, earl of Bothwell, a man under suspicion for conspiracy against the king. Armed with a commission to hunt down Bothwell and his party, a force of Huntly's men surrounded a house at Donibristle in Fife where Moray was staying on 7 February 1592 and called on the earl to surrender to them. The arrest attempt had a thin gloss of legal authority but the shadow of the feud lay underneath. Moray made a desperate bid to escape from the house burning around him only to be chased and murdered. The killer of Moray was executed for the wrongdoing, but popular feeling was roused against Huntly himself. He submitted to the king's authority, spent a week in custody and was then released to remain out of the public eye for some time. The slaying of the 'bonnie earl o Moray' entered the popular tradition and generated one of the great folksongs in the Scottish canon:

> Now wae be tae ye, Huntly!
> And wherefore did ye say?
> I bade ye bring him wi ye
> But forbade ye him tae slay.[8]

After a quiet period, Huntly resorted to avenging himself on his old enemies. He sent Cameron and MacDonnell allies to attack the Grants and Mackintoshes, and made moves against the earl of Atholl, who had been an ally with Moray in 1590. The dead Moray's brother-in-law Archibald Campbell, earl of Argyll, also came in on the fight, scenting opportunity to extend Campbell power into the north. A charge against Huntly that he had been corresponding with the king of Spain heralded a long and fitful fall from grace. A reluctant James VI threatened to outlaw him; the church tried to excommunicate him. Finally for refusal to submit to royal authority he and other Catholic nobles were declared rebels. On the braes of Glenlivet on 3 October 1594, Huntly and his allies defeated a force led by the earl of Argyll against them but declined to follow up their victory by confronting the main royal army. Strathbogie was destroyed. In the spring of 1595 Huntly went into exile but only for a year or so before he was once again in the north-east and, amazingly, thanks to some skilful diplomacy, back in James VI's good books. In 1599 he was appointed co-lieutenant and justiciar for the Highlands with his brother-in-law, the duke of Lennox. He joined the Privy Council and was raised to the rank of marquess. There were more feuds and conspiracies to come in the career of George Gordon before he died in 1636, to be succeeded by his son, another George, who fought and lost his head as a royalist in the Civil Wars.

During the few quieter phases in his life, when he had confined himself to his base in Strathbogie, Huntly had devoted time to building and strengthening the castles in Gordon hands. Other lairds were engaged on similar tasks, and the late sixteenth and early seventeenth centuries are one of the great castle-building periods, for both the construction of new castles and tower houses and the enlargement of existing strongholds. This was partly a fashion and declaration of status, especially for those who had acquired church lands during the Reformation, but it was also a response to troubled times and to the spreading use of firearms. A man of property could sleep more soundly if his outside walls were as high and as thick as he could make them. The boom in building must have provided welcome work for masons, carpenters, carters and other tradesmen in an otherwise bleak period. There were 24 periods of famine between 1550 and 1600, trials made worse by the fact that the population had been expanding faster than the economy. The Privy Council in Edinburgh issued repeated orders to ban or reduce the export of grain, the staple food – in June 1572 it even told citizens in the capital to foist themselves on their friends in the country where they might be better provided for – and tried to restrict travel and quarantine seafarers to limit the spread of infection.[9] The Chronicle of Aberdeen notes in 1578 'a great dearth of all kind of victuals through all Scotland, that the like was not seen in no man's day before. The meal was sold for six s[hillings] the peck, the ale for tenpence the pint, the wine from the best shipment forty pence the pint; fish and flesh were scant and dear'.[10] As prices rose, begging, vagrancy and emigration to the Continent either for trade or soldiering increased. Perhaps the economic situation was not as

Brodie Castle. The core of the castle is a Z-plan tower dating from 1567 but it has been extended and remodelled on several occasions over the centuries. The Brodie family claims a presence in the area from the eleventh century.

severe around the Moray Firth as it was in the central belt, a difference that may in part explain why contemporary writers refer to the fertility of the Moray Firth lands as if they are drawing a contrast between the cornfields of the north and the pitiful acres in the south. In the 1590s grain was being shipped from Caithness to Leith, the beginnings of what would expand into a major feature of the northern economy.

The tensions and stress in society may also have contributed to the outbreak of witchcraft trials, perhaps the aspect of the life of the period that is most puzzling to the modern mind. It was as if the intensity of religious uncertainty was channelled into confused and aberrant outlets – there had always been witches and spirits but now they were seen as agents of the devil, if not Auld Nick himself, creatures to be literally burned and sent to Hell before they could drag good Christians with them into everlasting torment. Accusations and trials peaked during the years between 1590 and 1662 in Scotland, at the same time as a second wave of persecutions across northern Europe. Most trials took place in the Lowlands of Scotland but some occurred in an arc around the north-east to Moray and the Black Isle. There were few in Inverness-shire, Easter Ross, Caithness and Sutherland, and hardly any at all in the Gaelic-speaking Highlands.[11] It was therefore ironic that the last execution of a witch in the country took place at

The fully restored Ballone Castle on the coast of Easter Ross gleams in the sun. The original was built by the Dunbar family of Easter Tarbat in the mid-1500s.

Dornoch in 1727. Belief in witchcraft did not cease. Many of the taboos and practices observed by fishermen almost certainly derive from witchcraft, and small, often private rituals associated with farming have survived into the present.[12]

An attempt to counter the influence of the Gordons in maintaining the Roman church was made by Charles Ferme or Fairholme, the new minister in Fraserburgh from 1600, with the backing of the laird of Philorth. 'In that Country, Popery was not yet much rooted out, and ther was great need of a learned ministry and a carefull education of the youth of the nobility . . .', noted the chronicler of church history, Robert Wodrow.[13] A college was established in Fraserburgh in 1599 to further the Protestant cause but the zealous Ferme did not stop there. In 1605 he and his fellow divines in the Synod of Aberdeen launched excommunication proceedings against Huntly, now elevated to the rank of marquis. This did not worry the latter at all but the Privy Council was outraged, issued letters of horning against the Synod and threw the ministers into prison. Six were banished from the kingdom after languishing in Blacknes Castle for a while. Ferme and John Monroe, minister of Tain, were confined and then sent into a form of

internal exile. 'At the proposal and desire of our Scots bishops' the clerics found themselves sent to 'desolate barbarous places of the Highlands where they wer under great straits and understood not the language spoken'. Ferme was posted to Bute, James Greg to Caithness, Nathaniel English to Sutherland, clearly all districts seen in Edinburgh as eminently arduous and remote enough for punishing errant ministers. The reference to not understanding the language spoken must be an exaggeration, at least in the case of Caithness at this time and probably also in Shetland, the place of exile for William Forbes. A few had repented or had been 'prevailed' upon to do so by the summer of 1607. There was no early release, though, for Ferme from Bute or for John Monroe from Kintyre. Early in 1608, Ferme wrote, 'I have to this hour been relieved by the comfort of no creature . . . A thousand deaths has my soul tasted of, but still the truth and mercy of the Lord have succoured me.' Ferme was finally freed but his sojourn in the isolation of Bute may have affected his health as he died in September 1617, at around the age of 50. The college he had attempted to foster in Fraserburgh had already closed.

The sixteenth century was more than a time of power struggle and religious ferment. It was also when enquiring minds under the influence

Left: Forres – the witches' stone. This boulder, set in the ground outside the present police station in the town, marks the site where a witch was burned. It was the custom, according to local tradition, that the alleged witch was placed in a barrel through which spikes were driven, rolled from the top of nearby Cluny Hill and then executed where the barrel came to rest.

Right: The date carved on this stone in Dornoch is wrong – it was in 1727 that Janet Horn was executed on this spot, the last judicially condemned witch in the country.

of the Renaissance began to look at the country as a political and cultural entity. George Buchanan, Latin scholar and poet, propagandist for the Protestant faith, wrote a new history of Scotland and dedicated it to his most important pupil, the young James VI. In his description of the Moray Firth region, he repeated much of what Hector Boece had said, mentioning the abundance of corn pasturage and orchards in Moray, the contrasting uncultivated mountainous regions and fertile plains in Ross, and the devotion of the Sutherland people more to pasturage than tillage. He repeats the curious fact that Loch Ness never freezes over – that several writers thought this noteworthy is possibly an indication of how cold winters were in the sixteenth century – and describes the safe haven provided for even the largest fleets by the Cromarty Firth once the 'stupendous cliffs' of the Sutors were bypassed. The earliest written information of how seafarers saw the Firth survives from 1540, when Alexander Lindsay, about whom we know next to nothing, compiled a rutter or pilot book for the voyage undertaken by James V to the 'the north and south isles for the ordouring of thame in justice and gude policy'.[14] For the Moray Firth, the rutter gives information about tides, courses, distances in 'kenningis' and miles, havens and dangerous rocks and shoals. For example, to reach Torrisnes [Tarbat Ness] from Buchan Ness one steered north-west; 'from the foreland of Murray to Invernes west southwest'. The 'Fyrth of Crumberte' [Cromarty] is praised 'above all hauens in the yle of Britane' as a safe anchorage in all winds.

One very significant innovation was Timothy Pont's attempt to capture the shape of the land on paper, pinning the important features – the mountains, the woods, the rivers, the settlements – in ink. What motivated Pont to do this remains a matter of great speculation but map-making had been going on in neighbouring countries for some time and perhaps that was enough to encourage such activity in Scotland. Unfortunately Timothy kept neither diary nor journal although his manuscript maps and some field notes survive. His father, Robert Pont, was a prominent figure – he was Moderator of the General Assembly five times and held other public offices – and in 1593 went on an extensive tour as a church commissioner in the far north. Timothy may well have accompanied his father and this is when he may have done the work for a commission to search for minerals and metals in Orkney and Shetland, prospecting that may have led to more general cartography throughout the rest of the 1590s. He was appointed minister of Dunnet parish in Caithness in 1601. His brother Zachary was given another Caithness parish, Bower, at around the same time. The last documented reference to Timothy is in a bond recording a loan of 1,000 merks by him to George Sinclair, earl of Caithness, at the end of May 1611. By 1614 he had been replaced by another minister in Dunnet. A considerable number of Pont's maps, some revised by Robert Gordon of Straloch, were eventually published in Amsterdam in 1654 in Johan Blaeu's *Atlas Novus*.

Feuds persisted into the seventeenth century as an ingrained custom probably exacerbated by the economic conditions. Few prominent families seemed to escape involvement. The several branches of the Dunbar kindred in Moray fought among themselves in what was tantamount to gang warfare, a continual source of unrest in the streets of Elgin and Forres; the sheriff was accidentally shot and killed in one melee in 1611. In Banff James Ogilvie of Acheeries was murdered in the street by his cousin, Sir George Ogilvie, the provost, in another kindred feud in 1628. A long-lasting grievance among the Mackintoshes over how they had been dispossessed by the earl of Moray found violent outlet in 1624 after the death of Angus McIntosh of Alturlie. Angus had, in the words of John Spalding, an Aberdeen clerk who kept a journal of his life and times, held 'his kyn and freindis of Clanchattan . . . under reull and in peace by his pouer and pollicie' but the young chief who succeeded him had no control over his followers and about 200 of them launched themselves against the tenants of Moray, attacking and pillaging their farms. Moray recruited 300 fighters, mostly McGregors, according to Spalding, from Menteith and Balquhidder but all they did was sit around Inverness at the expense of the earl when they should have been harrying the Mackintoshes into submission.[15] Moray sought the ear of the king, now enjoying his new surroundings in London, and asked to be appointed lieutenant in the north with legal powers to deal with his enemies. James readily granted the earl his wish, but this upset the marquis of Huntly who felt that he should have the lieutenancy himself, despite the fact that Moray was married to his own daughter.

Spalding describes many of the raids in vigorous Scots. For example, in 1634, 'thair brak out a number of Hieland lounis and heiryit the brayis of Morray, the victll deir at xvi [16] merkis the boll'. On that occasion, one of the pillagers was caught by a former provost of Elgin and duly convicted and hanged 'quhilk effrayit the rest of the lymmaris fra thair robbrie and oppressioun'.[16] Almost all the raids were aimed at obtaining livestock and

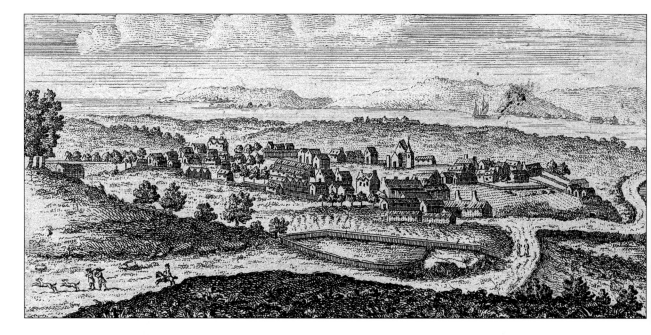

The Chanonry of
Ross, or Fortrose as
it is now known,
viewed from the hill
on the north side in
this drawing made in
the late 1600s.

grain. 'Upone the 15th of November [1634] thir Gordouns raisit out of the
ground of Frendracht about 13 scoir of nolt [cattle] and auchtein scoir of
scheip . . . Upone the 23rd of November thay brynt up the corne yaird of
the Maynes of Frendracht quhairin thair was standing four-score stakis.'[17]
For Frendraught to lose in the space of a week, and a week on the threshold
of winter to boot, 260 cattle, 360 sheep and 80 cornstacks must have been
economically crippling. He sought redress through the Privy Council but
when the government's herald, en route to Elgin to proclaim the summons
against the raiders, fell in with the Gordons, the perpetrators of the crimes,
he was told that they had been driven to it to seek revenge on Frendraught.
'The herauld glad of this ansuer and blithe to wyn away with his lyff
took his leve . . .', relates Spalding. Huntly and Frendraught reached a
reconciliation a year later. In the course of his account of this feud, Spalding
notes how men wanted for crimes fled across the Firth from Covesea to
Ross and Caithness.

After the Reformation the contest between the throne and the pulpit
dominated the nation's politics. James I and VI played his hand very well
in this game. In 1618 he managed to persuade the General Assembly
to accept the Five Articles of Perth to introduce a number of liturgical
practices, including kneeling to receive communion and the rite of

confirmation by bishops, although to many Presbyterians such customs had no scriptural authority and smacked too much of Rome. Scotland was now governed through the Privy Council meeting in Edinburgh but in constant communication with the distant monarch. After James's death in 1625, Charles I continued to rule in this manner but he lacked his father's knowledge of the fractious kingdom from which he had sprung, especially when it came to dealing with the church. The Act of Revocation in 1626 threatened to deprive the nobles and lairds of the church properties they had managed to secure during the Reformation. On his visit to Edinburgh in 1633, Charles further antagonised the nobility with his introduction of high-church ritual to Holyrood but his final and most serious mistake, affecting every rank in society, was a clumsy attempt to impose the use of a new prayer book. Reactions varied but were usually hostile. In October 1637 at the provincial assembly of the church in Elgin, the bishop commanded the ministers in his jurisdiction to use the new prayer book but some refused. A royal proclamation in December of that year ordering the prayer book to be adopted by Easter was too much. In February 1638 a document of protest, the National Covenant, the name a deliberate echo of the Old Testament, was drawn up and published for signature. The first mass signing took place on 28 February in the Greyfriars kirkyard in Edinburgh. The Covenanter movement was born. Copies of this manifesto-cum-declaration of principle were quickly sent out to the shires. It was April before it reached the Moray Firth but the news of its approach may have encouraged the action of 'certane scolleris' in Fortrose who in March 'cam pertlie into the kirk' and confiscated the offending prayer books, which the bishop of Ross had been using peaceably in the Chanonry kirk for some time, with a view to burning them. When a sudden shower of rain extinguished their match, they tore up the books and threw the pages into the sea. The bishop cut short his sermon and left for the other side of the Firth as quickly as possible where he conferred with the bishop of Moray and Huntly before travelling on south incognito to inform the king.

The earl of Sutherland, Lord Lovat, Lord Reay, chief of the Mackays, Lord John Sinclair, son of the earl of Caithness and the laird of Balnagown reached Inverness on 25 April with the Covenant for local signatures. Some of these northern leaders may have been showing an appreciation of political realities rather than principle. John Gordon, the thirteenth earl of Sutherland, was to navigate his way with considerable skill through the shifting political landscape. The lands of Sutherland had been erected into

Fortrose Cathedral, originally constructed in the late thirteenth century, became a victim of the Reformation like its counterpart in Elgin (p. 106). Removal of lead from the roof in 1572 and of building materials at various times after that sealed the structure's fate. Euphemia of Ross, widow of Walter Leslie and then of the Wolf of Badenoch, has her tomb here. The Mackenzies of Seaforth continued to use the cathedral as a burial place into the nineteenth century.

a sheriffdom separate from Inverness in 1631, and Dornoch had become the county town. Donald Mackay, created Lord Reay in 1628, was a royalist who yet signed the Covenant. James Stewart, the fourth earl of Moray, was also a royalist but wisely decided to stay out of things altogether, spending much of his time in England. The combined presence of the nobles must have impressed the Inverness authorities, for 'The haill toune, except Mr William Clogie minister . . . and sum few otheris, willinglie subscrivit,' noted Spalding. This pattern was repeated across the north – 'Caithness, Sutherland, Ross, Cromartie, Narne . . . for the most pairt subscrivit'. In Forres, the entire presbytery except the minister of Dallas signed. The Elgin minister John Gordon, perhaps unsurprisingly in view of his kinship, was the only one to refuse in that burgh. The bishop of Caithness, thought Spalding, was against the service book. The bishop of Moray, though, made ready to resist the Covenanters and stocked his palace at Spynie with supplies, powder and shot. Negotiations between king and kirk failed to reach agreement. In December 1638 the General Assembly re-established plain Presbyterianism and out went bishops, a defiance of royal authority that precipitated the first bishops' war. Early in March 1639, three Covenanters – the ministers of Rafford and Ardclach and the provost of Nairn – accosted the bishop of Moray at his church door, told him he was a bishop no longer and called on him to repent. 'The good bischop . . . left af to preiche ilk Sonday . . . and resolvis to keip his castell of Spynie cloiss, and cum no more out . . . bot he wes forsit to give it over or [before] all wes done,' noted John Spalding. In July 1639 the Presbytery of Elgin noted the bishop of Moray's excommunication, but he stayed on at Spynie until July 1640 when he was forced to move south to Edinburgh to the Tolbooth prison. Released in November 1641 on condition he stayed away from Moray, he died on his own lands at Guthrie in August 1649. Sir George Ogilvy in Banff refused to sign the Covenant, although the people in the town were for the most part in favour of it. 'Albeit in former tymes they depended much upon Banfe's family . . . yet now they wer so far estranged from him that they wer growne his enemyes,' noted James Gordon.[18] A doctor called Alexander Douglas was the local leader of the pro-Covenant majority, a stance that brought him a series of public offices – provost, sheriff, member of Parliament – in the future. Ogilvy's immediate fate was less pleasant – a Covenanter army camped on the Dowhaugh in August 1640, destroyed his orchards and woods and left nothing standing of his house save the bare walls.

What fighting there was took place mainly in the south. Forces sent by Huntly, as leader of the royalists in the north-east, to provision and garrison Inverness Castle were confronted and stopped by a mixed group of Frasers and Mackenzies: '[they] wold not suffer him [William Gordon] to enter and violentlie and maisterfullie reft and took fra the gentilman his haill armes and trunkis . . . The gentilman . . . wes blyth to wyn away.'[19] The Frasers then occupied the castle and, according to John Spalding, a royalist sympathiser, wrecked the furnishings and the library. Just after this incident there occurred another somewhat farcical encounter when Huntly and his followers accidentally found themselves in the vicinity of Turriff at the same time as Covenanter forces led by the Marquis of Montrose; each side forbore to attack the other. At every market cross around the Firth, Huntly was proclaimed the king's lieutenant in the north 'from the North Esk to Caithness' but this did not prevent the Covenanters rising in force. Spalding estimated them to have numbered around 9,000 when they assembled at Kintore under Montrose. Outnumbered and vulnerable in Strathbogie, Huntly thought of what he had to lose, opened negotiations – one of his lieutenants was Robert Gordon of Straloch who was to play a crucial role in bringing Timothy Pont's maps to the wider world – and, after some discussion, he and Montrose peaceably went their separate ways.

A few months later, talks at Berwick between Charles and the

Covenanters broke down. During these negotiations, however, Montrose who up until then had been a leading secular figure among the Covenanters began to change his mind about the extremist Presbyterians around him. He came to see them as betraying the original Covenant that he had signed in Greyfriars and finally resolved conflicting loyalties by going over to Charles, a long process that did not reach its fateful conclusion until August 1643, a year after full-scale civil war had erupted in the kingdom. Montrose, appointed royalist military commander in Scotland, brought the war to the shores of the Moray Firth. His exploits have been often recounted and a summary will suffice here. In the autumn of 1644, with his own troops augmented by a force of Irish of Clan Donald under Alasdair MacColla but still numbering only a meagre 2,000 or so, he cut a swathe across the Highlands, sacked Inveraray and the heartland of the Campbells, and inflicted a massive defeat on the pursuing forces of the marquis of Argyll at Inverlochy. He then marched past Inverness to Elgin, causing panic – 'the toune's people . . . fled also with thair wyves, barnis and best goodis . . . and few baid within the toun throw plane feir . . .', noted Spalding.[20] Montrose arrived two days later and accepted a payment of 4,000 merks to leave the town unburnt. The laird of Grant's men, who had joined the marquis, were less obliging and pillaged everything they could carry away. Montrose continued his march to the east. At Cullen he found that the earl of Findlater, a Covenanter, had fled to Edinburgh and left his wife at home. Lady Findlater had to watch all her furnishings, fabrics and silverware being taken by the soldiers before she 'pitifullie besocht' Montrose not to set fire to the house but to wait for 15 days to see if her husband would return. If Findlater stayed away, she said, Montrose could do as he liked. Her courage impressed the marquis who accepted her promise of 20,000 merks and left the house standing. In Banff the royalists stripped the clothes from anyone they met in the street, burned a few worthless houses but shed no blood.

A climax of a sort was reached in the early summer when Montrose's army and that of the Covenanters under Sir John Hurry met in battle at Auldearn. Hurry had a mixed force, his own troops augmented by clansmen: the earls of Sutherland and Seaforth with their armed followings, the earl of Findlater no doubt keen to avenge the previous slight on his honour, Frasers but not Lovat in person – he was now in his fifties and would die in the following December, and various others – Inneses, Rosses, Munros, Dunbars – in total some 4,000 foot and 500 cavalry. Spalding noted that Seaforth was suspected of treachery, as he had already sworn to Montrose

for the king's service and had now broken parole to join the Covenanters. On 8 May, Montrose pitched up in Auldearn, caught between Hurry's forces and a pursuing Covenanter army that, according to James Fraser in his contemporary chronicle 'was gotten betuixt the mountains and him'.[21] Resolving to spring a surprise attack, Hurry led his forces on the offensive on the following morning. The battle swung back and fore to the west of the village, the movement of the troops constrained by areas of marshy ground, but in the end Montrose's men prevailed. Spalding's version of the outcome is far from accurate but is worth quoting for the way it sees God's hand in the victory: 'Thair wes reknit to be slayne . . . above 2,000 men to Hurry and about sum 24 gentilmen hurt to Montrose, and some few Irishes killit, which is miraculous and onlie foughten with Godis awin finger . . . so mony to be murderit and cut doun upone the ane side and so few on the other . . .'.[22] James Fraser wrote that losses among the Frasers left 87 widows to harry Lovat with cries of 'this we got for our dissloyall, rebellious covenant which we fought for'.[23] Hurry withdrew to the west, and on the

The battle of Auldearn was fought in these fields on 8 May 1646. The town of Nairn lies in the distance.

131

following day Montrose led his men off on another campaign of plunder against Covenanting lairds. Some houses in Elgin were burnt. Unfortunately the burgh council minutes of Elgin are missing for this crucial period but entries in the burgh court book make some bald references to things lost by ordinary people. The royalist campaign ended in defeat in the autumn of 1645 in the Borders but Montrose escaped and eventually made his way to Strathbogie where he succeeded in persuading a reluctant Huntly to come out in the royalist cause. Several of the clans – Macdonalds, Macleods, Mackenzies and Mackays – had meanwhile reconsidered their adherence to the Covenant and were about to declare for the royalists.

By April 1646 Montrose was in a position to venture again for the king. His plan to attack Inverness failed when the Gordons instead of supporting him chose to indulge in some old-fashioned raiding on traditional enemies. According to James Fraser,

> [They] ravaged up and down the country upon all suspect persones, burning cornyards, and pillaging all townes and villages . . . putting taxes and stents upon the people . . . plunder of all sort fit for portage was carried away, so that all Murray betuixt Spey and Nesse was wasted . . . Sir Robert Gordon of Gordonstoun was a good friend to many about him and prevailed with Lord Lewis [Huntly's son] to spare the corns, especially in the parish[es] of Duffes, Aves and Cinedward. Some they favoured as their own . . . but the Inneses, Brodyes, Dubarrs, Kinards, nay the Stuarts and Rosses, were made a prey . . . The barnyards of Geddes, Brody, Calder, Clava, Kilravock and Castle Stuart were all burnt to ashes; and the very instant time in which Lord Hugh Fraser of Lovat were interd [near Beauly] his great cornyard at Dalcross was all in a fire to our view. Of this inhumane fact I myselfe was eye and hundereds witnesses at that solemn burial.[24]

The Covenanting forces in Inverness escaped to the Black Isle before Montrose could throw a cordon around the town. While his gunners set up their cannon to batter the castle, some of Montrose's cavalry and foot soldiers forded the Ness to harry the countryside beyond. James Fraser remembered vividly this foray through his parish: 'Betuixt the bridge end of Inverness and Gusachan, 26 miles [at the head of Strathglass], there was not left in my countrie a sheep to bleet, or a cock to crow day, nor a house unruffled'.[25] The Covenanting army under Major General John Middleton crossed from the Black Isle to Moray to outflank Montrose and approach from his rear, but again the marquis escaped, westward to Beauly. Middleton led a somewhat half-hearted chase but finally allowed the royalists to retreat. The General Assembly excommunicated the marquis and along with him his ally, the Mackenzie earl of Seaforth. A few months later, Montrose was ordered by Charles I to disband his army and go into exile.

The victory of the Covenanter and Parliamentary forces in the civil wars brought respite to the shores of the Firth. In September 1646 in Inverness, a committee listed the destruction wrought on the town on a roll of paper over 28 feet long, recording damages to goods or property, the costs of quartering troops, burnt crops, the destruction of one-third of the wooden bridge over the Ness, and so on – in all a total of £93,959 Scots. No records were made of the havoc inflicted on the rural peasantry but the account by James Fraser makes it plain that they suffered grievously. Some of the royalists now had to reconcile themselves to the Covenanters being in power. In Elgin, for example, the laird of Pluscarden confessed 'ingenuouslie with signs of great sorrow and humilitie', according to the presbytery records, how he had joined the excommunicate traitor James Graham.[26] The kirk triumphant always referred to Montrose as simply James Graham or as 'the public enemy'.

The peace did not last long. The execution of Charles I on 30 January 1649 shocked even some extreme Presbyterians in Scotland and in the north roused the royalist leaders to descend once more on Inverness; these clansmen were surprised and overcome by Parliamentary troops near Balvenie three months later. Montrose made a final effort in the royalist cause in 1650, landing in Orkney from Bergen, crossing to Caithness and marching south at the head of a small force only to be defeated at Carbisdale; Montrose himself went into hiding but was betrayed, captured and executed, in the process adding himself to the Scottish gallery of romantic lost causes. The mouth of the Spey witnessed in June 1650 the landing of the exiled Stuart king. His coronation as Charles II followed soon after in Scone. To accomplish his return, he had agreed to support the Protestant religion in Scotland, to become in effect a kind of constitutional monarch in a theocratic state. In England, where revolution had gone further than north of the Border and had installed Cromwell's Commonwealth in power, there was now fear of a Scots invasion, as had happened in 1648. The solution was a pre-emptive blow. In July 1650 Cromwell's army invaded, defeated the Scottish forces at Dunbar and gradually assumed dominance south of the Forth, before launching a further advance up the east coast through Fife. A Scots counter-invasion brought about the final confrontation in September 1651 at Worcester. Charles escaped from the battle into exile but many of his Scots followers were sentenced to slavery in New England and the West Indies.

General George Monck was given the task of subduing the remainder of Scotland. In his manse at Wardlaw, James Fraser wrote that the north was 'demure under a slavish calm'.[27] One of Monck's regiments, under the command of Colonel Thomas Fitch, passed through Aberdeen in November 1651. On 1 December, 'the Inglische armie cam . . . no lectour,' noted the Elgin kirk session, adding on the following day, 'No preaching, the armie being in the toun'.[28] In the event there were to be no readings or sermons until the sixteenth, while the kirk session held its breath: 'Disciplein is suspended a litle whyle [to] sie what the Inglisch armie doth'. The troops were dispersed in garrisons as far north as Caithness where they occupied Ackergill Tower and also established a presence in the parish of Canisbay on the southern shore of the Pentland Firth before, in February 1652, landing in Orkney. At first this was an unsettling military occupation. At the end of December, the Forres presbytery noted that 'the English Armie is ane enemie to Presbyteriall Government and would not fail to marke narrowlie our course and carriage therein'.[29] In time, and probably quite quickly, the English soldiers adjusted to their new environment and before long kirk discipline reasserted itself to lament many instances of the locals and incomers indulging in drinking, fornication and other evils. In May 1652 work began on a great new army citadel in Inverness, a strategic location at the mouth of the Great Glen from which to keep an eye on the royalist clans in the hills. The construction took four years and, in James Fraser's estimation, cost £80,000. The minister disapproved of the reuse of stone from various churches and religious buildings, among them Kinloss Abbey and Fortrose Cathedral, but had to admit that the citadel was 'a most statly sconce'. It provided employment for scores of labourers at a shilling a day, a boost to the local economy. Inverness enjoyed greatly increased trade, new products and commodities appeared in the market and, apparently for the first time, the town acquired a surgeon and an apothecary. 'They not onely civilised but enriched this place,' observed Fraser.[30] Government policy was to base mobile garrisons in such places as Dornoch, Ruthven in Badenoch, Inverness and Inverlochy as a barrier across the north to prevent any rebels from descending easily on the Lowlands. A naval vessel was stationed on Loch Ness. It may have seemed ironic therefore when what was to prove the one major insurrection began in the south in Atholl, in August 1653 under the leadership of William Cuningham, earl of Glencairn. After campaigning along the southern edge of the Highlands this armed force chose to move north to Aberdeenshire and Moray, and at the end of February 1654 was reinforced by contingents under General John Middleton who landed at the mouth of the Dornoch Firth from the Continent. Divisions among the command weakened the royalists – Glencairn and Sir George Munro even fought a duel at Evelix – and despite a few victories in skirmishes the rising eventually failed in a final clash with Cromwellian troops at Dalsnaspidal on the evening of 19 July. Middleton escaped into exile again with Charles II, and one by one the clan chiefs made their peace with the new regime.

The Commonwealth was an experiment before its time. That England and Scotland could be united in a single political entity, and a republic to boot, was a revolution too far for the mid seventeenth century, especially when led by such a man as Oliver Cromwell. At least James Fraser in Wardlaw thought so – in his history he lists the 'tyrannical proceedings' that the military authorities pressed on anyone of a royalist persuasion.[31] Unlike in Ireland, however, Cromwell's name evokes little emotion in Scotland. After he died in 1658, his son Richard succeeded him but it soon became clear to General Monck and the army that the son was not from the same mould as his father and, in what was essentially a military coup, the exiled Charles II was restored to the throne in 1660. According to James Fraser, 'In Scotland there was never greater joy and acclamations heard in every honest mans mouth and heart'.[32] The imminent prospect of the withdrawal of the English garrison from its citadel on the edge of the town – it left in April 1662 – worried the burgh council of Inverness, fears they recorded in the council minutes and which they allayed by forming a night watch to patrol the streets. Charles II's reinstatement of the episcopacy and attempts to impose some uniformity in church worship provoked the reaction of the more extreme Presbyterian factions, the Covenanters, but their struggle was largely confined to southern parts of Scotland.

In the Highlands the clans remained largely royalist and in April 1689 some of them had an opportunity to demonstrate this when John Graham of Claverhouse, 'Bonnie Dundee', raised the standard of rebellion in support of James II and VII who the year before had gone into exile after being replaced on the throne by William of Orange. In search of recruits and support in this the first outing of the Jacobites, Dundee came to Huntly's territory in the north-east and then travelled through Keith to Forres before turning south again, with government troops led by Major General Hugh Mackay of Scourie in pursuit. The rising turned into dangerous farce when the MacDonalds of Keppoch came out in support of Dundee and invested Inverness, demanding 4,000 merks and a scarlet lace coat for the chief, on pain of putting the burgh to the torch. Dundee managed to persuade the clansmen to accept what the frightened burgh could offer but then they abandoned him and went home to Lochaber. With only some 200 followers, Dundee evaded his enemy by moving south through the Great Glen away from the Firth lowlands. The opposing forces finally met in the pass of Killiecrankie on 27 July where Dundee was victorious but died from a fatal wound in his hour of triumph. The rising dwindled over the winter but displayed a brief resurgence in the following year before a final victory for the government in May at the Haughs of Cromdale.

The last decade of the seventeenth century saw the worst famine within contemporary memory, a period recalled as 'King William's ill years'. The extent of this catastrophe has only begun to be explored, a surprising oversight in view of its impact. It is estimated that Aberdeenshire lost about one-fifth of its population, with a smaller but also significant fall in the numbers in Angus and Perthshire.[33] Harvest failure and starvation were widespread, along with epidemics and breakdown in social order. The evidence for conditions around the Firth is sparse, but there are telling references in surviving records. In August 1698 the parish kirk collection in Inverness was designated for the poor in 'this hard and strick tymes of ffamine'.[34] Seven out of the eight harvests between 1693 and 1700, the severest cold period during the so-called 'little ice age', failed in all the upland parishes. Grain was imported from the Baltic. A peak in mortality is very clear in the records for the parishes of Kirkhill and Kilmorack for the years 1695–1700, and there is a matching fall in baptisms in Dingwall and other parishes.[35]

The Firth lands suffered from being the arena for the last extended land warfare in the British Isles, the Jacobite rising in 1745 aimed at restoring the ousted Stuart dynasty to the throne. Previous Jacobite attempts had had less effect locally. The first came hard on the signing of the Treaty of Union in

1707. In the following year a French fleet was prevented by the Royal Navy from effecting a landing of Jacobite forces in the Firth of Forth and moved north to make a second attempt to put men ashore near Inverness, but this plan also fell through when contrary winds kept the ships from approaching close enough to shore. After some dodging to and fro the French gave up and sailed for home, a case of the weather saving the day. The accession of George I in 1714 provoked some disturbance in Inverness, pro-Hanoverian sermons from the pulpit pitched against magistrates and a provost who were decidedly pro-Jacobite. When the earl of Mar launched his rising in September 1715, Mackintoshes seized the town and amused themselves by looting pro-Hanoverians before setting off south, to be replaced by a pro-Jacobite garrison of Mackenzies who in turn were overcome when local pro-Hanoverians rallied and retook control. In 1719 the Jacobites made a third foray, landing a force in Loch Alsh from Spanish ships; this effort came to a dismal end in a clash with troops amid the heather and rocks of

In the foreground of this depiction of the Battle of Culloden, someone seems to be attempting to lead Prince Charles Edward Stuart to safety as the Jacobite army is cut to pieces by the government troops.

137

The army base of Fort George was built at the tip of the Ardersier peninsula after the victory of the British government in the last of the Jacobite risings in 1745–46. It controls the narrow gap between itself and Chanonry Point, the entrance to the Inner Moray Firth and the port of Inverness. In the centre of this picture, Loch Ness lying in the trough of the Great Glen is plainly visible.

Glenshiel. By the time of the last Jacobite rising in 1745, the majority of the population appear to have been content and enjoying growing trade under the Hanoverian regime. In the rural hinterland, however, many of the clans remained sympathetic to the Jacobites and rallied to the colours of Prince Charles Edward Stuart when he arrived in the Highlands. The story of the rising is well known – the march south, the occupation of Edinburgh, the thrust into England that ended with the turning back at Derby, the retreat to Inverness in the late winter.[36] The response of the communities around the Firth to the Jacobite cause showed a society polarised. The muster rolls of the Jacobite regiments include men from every part of the country but participants in significant numbers came from particular localities. Lord Pitsligo's regiment of horse was raised in the north-east, Chisholms, Frasers, Gordons and Mackintoshes came from their traditional lands, and the earl of Cromartie's regiment recruited mainly from the inner Firth coastlands. Most of Caithness and Sutherland adhered to the status quo –

Cromartie's men were defeated at Little Ferry by local militias on the day before the final battle at Culloden, as did most townspeople and a large proportion of the general population who probably wished simply to see the back of any army. Elgin, Forres and Inverness were occupied by the combatants but the last suffered the most from the initially hostile and frequently brutal presence of Cumberland's forces. In a sense the Jacobite risings were the last spasms of a dying order and already by the time the blood flowed on Drumossie Moor other currents were at work that would power a transformation of the country around the Firth.

Fort George, as seen in the 1780s from the north side of the Firth.

# CHAPTER 10
# CHANGING THE FACE OF THE LAND

Braco's Banking
House in Elgin,
sketched in the early
1800s along with a
passing angler and
his dog, is now home
to a cycle shop.

Braco's Banking House is the curious name for one of the fine old buildings in Elgin's High Street. The crow-step gables and the arches speak of age, an impression confirmed by the date on the dormer windows – 1684. The initials of the couple for whom the house was built – John Duncan and Margaret Innes – are also carved in the stone, as was the custom, but their memory is overshadowed by that of the man who bought the place in 1703 and gave his name to it – Alexander Duff of Braco. The Duffs displayed an ability to acquire wealth, and with it importance and enemies. An observer grumbled they 'all abounded in money', with an eye for the good opportunity when land was cheap, interest high, rents low and other

families strapped and sunk in debt 'by means of the Civil wars and other public commotions'. Their purchases included the estate of Balvenie in 1687. Stories were told about them. One tale has Alexander Duff eyeing the smoking chimneys of a township in Mortlach and saying 'I'll gar a that reek gae through ae lum yet'; and another has the earl of Kintore praying to his maker to keep the Hill of Foudland between him and the Duffs. Alexander was member of the Scottish Parliament for Banff and against any union with England to the extent that he once threatened to behead a wavering colleague like an onion.[1] In September 1700 he arrested two notorious freebooters James Macpherson and Peter Brown at the Summerrive Fair in Keith. Macpherson enjoyed the protection of the laird of Grant who, on learning of his man's arrest, gathered

a band of followers, cornered Alexander Duff and forced him to let the robbers go free. Duff promptly recruited two fellow justices and sixty men and rearrested Macpherson and Brown. James Macpherson was eventually hanged at Banff where, according to the tradition that has made him into a Robin Hood figure, his last act before the noose did its fatal work was to play his fiddle and break it across his knee. Alexander Duff died in 1705. His son William did not long outlive him – taking his own life while in exile after the 1715 Jacobite rising – and the family wealth was inherited by Alexander's younger brother William Duff, who had already made a fortune as a merchant in Inverness and had bought land in Moray, including the estate of Dipple in 1684 for £70,000 Scots. William of Dipple died in 1722, leaving to his various offspring and friends a total of £50,000 Scots, roughly equivalent to £570,000 today. This William's son, also called William, was made Lord Braco in 1735 and the earl of Fife in the Irish peerage in 1759; more importantly for our story, he commissioned James Gibbs to design a new mansion at Balvenie and William Adam to create Duff House.

Several other buildings in Elgin have distinctive arched facades, such as this one at 42–46 High Street, thought to be the location of the Red Lion inn where Dr Samuel Johnson had an inedible meal in 1773. The arch on the left leads through a pend to a back court.

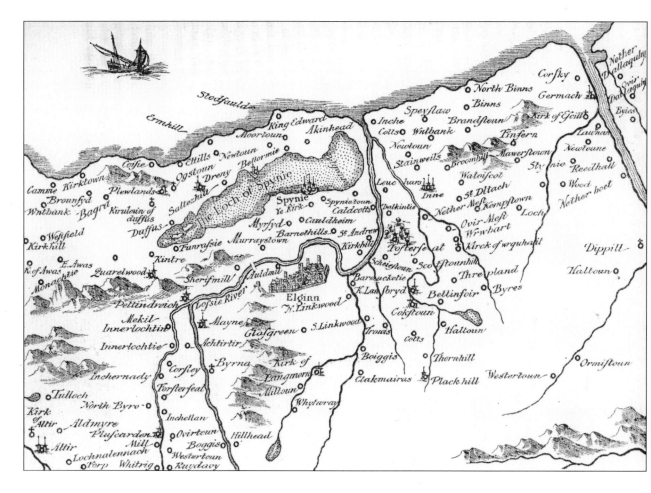

The lowlands of
Moray in 1662,
showing the extent
of the Loch of Spynie

In the early 1700s, rural tenants paid their rents mostly in kind, as had been the custom for centuries. There was very little hard currency in circulation, perhaps only some £800,000 in the whole of Scotland, half of which was invested and lost in the Darien scheme.[2] Families with cash like the Duffs provided the banking services and almost all such families were the larger merchants to be found in the burghs. They operated wholesale trade, shipping goods to other parts of Britain and across the North Sea, and at home supplying all manner of necessities and luxuries to whomever could afford them and often to lairds whose consumption of goods outran their credit. An account book of Inverness merchant William Dallas survives from the 1680s and lists his sale of tobacco, sugar, prunes, spices, paper, silk, serge and dyes to his customers in and around the town.[3]

The lairds of Mey in Caithness were conspicuous consumers in the early 1700s of fruits, spices, fabrics and other luxuries, running up long accounts with merchants not only in Thurso and Wick but also across the Firth in Elgin.[4] The correspondence of the merchants who worked in seaborne trade discusses cargoes of salmon, herring, iron, oatmeal, timber, wine; their contacts range along the whole Atlantic seaboard of the Continent and east to the far Baltic shore, as well as the ports and havens around the home coast. The risks were high but so were the rewards when a ship hoved homeward from the Firth mist.

The trading monopoly enshrined in the charters of royal burghs became increasingly circumvented and eroded. From the fifteenth century onward, landowners followed the precedent set by the crown and applied for the

A view of Elgin in the 1690s. The town is seen from the north side, with the main part lying between the castle mound, called Ladyhill today, on the west and the cathedral and its associated buildings to the east. The Lossie winds across the middle distance.

The Mercat Cross
and Tolbooth lend
dignity to High Street
in Forres. Erected
in 1844 to a design
by the Edinburgh
architect Thomas
Mackenzie, the Cross
with its pinnacles
and medieval-like
heads looks a bit
like a mini-Scott
Monument. The
Tolbooth, built in
1838 by William
Robertson, replaced
its forerunner on the
same site.The pattern
of streets and lanes
in several burghs
preserves much of
the original medieval
street plans.

right to set up burghs, so-called burghs of barony, with their own trading rights. Leading noblemen could also found burghs of regality in a similar fashion. Fraserburgh, under its original name of Faithlie, began as a burgh of barony before being raised in 1601 to the status of a burgh of regality by Alexander Fraser of Philorth and renamed. The possession of a charter did not ensure economic prosperity and many burghs of barony failed to grow much beyond the size of a village, while others became sizeable towns.

The commodity that figured most prominently among the goods traded by the merchants around the Firth was usually referred to as 'victual' – a collective term for grain, the staff of life, predominantly oats and barley, either as meal or as whole grain. The trade was already well established in the seventeenth century and continued during the following one, apart from a major upset in 1782 which was 'very remarkable for the lateness and coldness of the harvest'.[5] Cutting the grain in Birnie parish in Moray could not start until mid October and about half the crop was damaged by frost and snow. The result was a temporary reversal of the direction of the trade:

for a time the communities around the Firth were importing around 100,000 bolls of corn and meal each year but, by 1793, that had come to an end and exports from Findhorn, Lossie and Garmouth reached 10,000 bolls,[6] approaching the quantities normally sent out from the district. The victual trade was underpinned by the runrig system of agriculture, the traditional form of land distribution and cultivation whereby a farm was worked by a number of tenants in a communal manner, the families living in the fermtoun, tilling the 'infield' in strips of ground, the rigs, that they rotated among themselves year by year, and keeping livestock on the pasture of the 'outfield' and common grazings. The origins of this system, which was almost standard throughout the Scottish Lowlands by the mid 1700s, remain lost in the Middle Ages; the terms of runrig, infield and outfield first appear in records in the fifteenth century. The landlords received every year much of their rent in victual, which they in turn sold in the Lowland or Continental markets.

This established mode of life was swept away as an agricultural revolution spread northwards. A major influence driving this implementation of new ideas was the Society of Improvers in the Knowledge of Agriculture in Scotland, a group of landowners who began to meet in Edinburgh in 1723. Reputed to be the first group of its kind in Europe, the society lasted for around 20 years but left its name in the standard label for the years that followed – the age of Improvement. Many of the members were politicians and lawyers, who liked what they saw on their travels in England and the Continent and were in a position to innovate. At first not a great deal of capital may have been required, as at least some of the labouring to create

The Steeple in Banff's Low Street was erected in 1764 by John Adam and mason John Marr with a very distinctive spire. Originally free-standing, it has been attached to the adjoining Town House since 1796.

dykes and drains – two crucial tasks in improvement – could be done by tenants on the 'service days' that were part of their rent. New seed, implements and stock had however to be purchased and could prove costly. John Cockburn of Ormiston started his improvement in 1714 but he finally went bankrupt and had to sell his estate in 1747. One suspects an element of local pride in the comment of the minister of Boyndie in 1797 that 'This parish was one of the first in the north of Scotland in which the new husbandry was attempted and carried on with success' but there was also a strong degree of truth in it.[7] It was here, on the eastern corner of the Firth region, that the earl of Findlater in around 1754 implemented improvement on his

The rolling plain of central Caithness, an area that produced crops of oats and barley in the eighteenth century for export, is still heavily farmed for livestock and grain.

farm of Craigholes, and brought in an English overseer versed in the new ways. James Ogilvy, the sixth man to bear the title of earl of Findlater, had his keenness for improvement sharpened during his travels as a young man and, before he inherited his father's title, he had sent sons of his tenants south to learn the new agriculture. He is also credited with the introduction of the turnip to the north-east. The gentlemen farmers formed clubs, debating at their regular meetings the merits of lime, whin dykes, crop rotations and new ploughshares. Among the first was the Small Society of Farmers in Buchan, founded in 1730 but short-lived. Banffshire Farming Society was founded in 1785, two years after the Highland and Agricultural Society was formed in Edinburgh, the Ross-shire Farmers' Club in 1794, the Nairnshire Farming Society in 1798, the Easter Ross Farmers' Club in 1811 and the Caithness Agricultural Society in 1830.

The essential method adopted by the landowners was to form large farms from several smaller ones and offer these farms on long leases to single tenants instead of to communities. Findlater gave his tenants in Boyndie leases of 38 years and more. Rents which under the runrig system were traditionally paid in goods and service with only a proportion in money were now being converted into straight cash payments. The landowner benefited from a higher rent, the single tenant from having sole occupancy of a large area with time to experiment and recoup a profit. Some of the former small tenants found new roles as farm servants but many were simply shoved aside, given notice to move off the land their families may have occupied and worked for several generations. By 1796 the parish of Botriphnie in Banffshire had suffered from a drop in population as the rural poor, unable to afford the rising rents, flitted to other parishes where villages and towns offered some opportunity. In Lhanbryde fermtouns that had been home to ten to 16 families in the 1760s were, by the 1790s, let to one man, in some cases one from Elgin, in others one from another parish 'some of whom have no resident servant and others only one or two, where numerous families dwelt'.[8] The 'union of farms' was frequent in Cromdale, where 'one man now occupies as much land as was, 40 years ago, possessed by five or six families'.[9] This probably did not happen everywhere as, in Alves in Moray, the minister noted in 1794 many farms of the earl of Moray continuing in the possession of the same families for generations, with one family called Anderson able to trace their tenancy back for more than 400 years. The revolution was clearly a piecemeal one, proceeding fitfully at first and then faster towards the end of the century as rising rents and prices, pushed up by the wars with France, brought in new funds for investment. It was not happening fast enough for the parish minister of Wick in 1793: 'A pernicious custom still too much prevails in this and other places,' he wrote in reference to runrig, or 'rig and rennal', as it was termed in Caithness. 'This is necessarily attended with confusion and disputes, and is a practice requiring to be abolished'.[10]

An impetus was given to the revolution from an unlikely quarter. The lairds and nobles who declared their hand in support of the Jacobite cause lost their lands to the government, and in 1715 the Westminster Parliament appointed a group of commissioners to take over the management of these forfeited acres. Few of the later commentators on this initiative ventured as boldly into print as A.H. Millar did in 1909 in the introduction to his *Selection of Scottish Forfeited Estates Papers* for the Scottish History Society, where he speaks of the 'great impetus given to civilisation by the wise administration of the Commissioners',[11] but there is no doubt that the factors and administrators on the forfeited estates set in train the latest ideas on rural economic development. There were two phases of forfeiture – after 1715 and 1746. No estates bordering on the Moray Firth were affected in the first phase and only a handful, including the important Lovat Fraser

Grain was stored and rents were paid in this storehouse or girnal at Foulis on the Cromarty Firth. It was built in about 1740 for the local estate. A ferry used to run from here to the Black Isle.

estates and those of the Macleod earls of Cromartie, in the second but the influence of the commissioners' work extended everywhere. By the Disannexing Act of 1784, estates still in government hands were restored to the heirs of their former owners, and the proceeds from the sales, rents and other income were made available to fund public and charitable works such as the Society for the Promotion of Christian Knowledge (SPCK) and the schemes of the British Fisheries Society, all of which were to have their own effects on the Firth region.

In 1755 the commissioners gave the task of surveying the Lovat and Cromartie lands to Peter May, offering him a salary of 13 shillings per day from which he was expected to pay five assistants. Born around 1730 in either Aberdeenshire or Moray – his father belonged to Turriff – May may have learned mathematics, basic engineering, agricultural science, surveying and map-making under John Forbes, the estate factor at Kildrummy and in 1745 in the employ of the earl of Findlater. He certainly learned well for, at the time of his appointment, he was described as 'the best land surveyor in Scotland'.[12] May, in his turn, trained others, creating the small class of professionals who shaped the transformation of the countryside in the late eighteenth century. On 10 November 1755, from Castle Leod near Dingwall, May wrote to Lord Deskford: 'I came to this country the first of August . . . the tenantry here are the poorest people I ever saw; they have neither meat nor clothes. The reason of this I apprehend is partly owing to the small possessions they occupy, partly to the oppressive way they have been used by the landlords, but most of all by their own indolence and laziness.' This was a harsh judgement but one that we find repeated over ensuing decades as the apostles of the new agriculture moved around the head of the Firth and further north into the territories of the clans. May also thought the tenants in Strathpeffer distilled, brewed and drank too much. He noticed the sub-tenants, the cottars or 'mealers', as they are often called in a misspelling of mailer, meaning renter, struggling to subsist on small patches of ground held from larger tenants. In the surveyor's letters the chasm in attitudes and manners between the clanspeople and the lowlanders yawns wide. It is significant that of New Tarbat, the Cromartie estate in Easter Ross, May wrote: 'The tenants here are in a better way than the barony of Castle Leod, though I don't think them well in comparison with Banffshire' – the Firthlands offered contrasting stages in the agricultural revolution, with the eastern littoral more similar to Moray in character than the upland districts. At the end of October 1756, May began to survey the Aird of Lovat. A decade later we find him dividing the runrig lands of Lovat and Cromartie into farms, laying out settlements for disbanded soldiers. Soon after that task, May moved back to the east to live in Elgin and be factor on the earl of Findlater's Moray estates.

Various strands of activity unrolled as the countryside changed and as different factors came into play. We glimpse in the accounts of each parish written for Sir John Sinclair's monumental *Statistical Account of Scotland* (OSA), published in 21 volumes in the 1790s, how the patterns shifted at different rates and in varying ways across the land, depending on conditions and the reactions or enthusiasms of key local figures. According to the minister of Elgin, it had been Sir Robert Gordon of Gordonstoun who, in about 1765, had brought in the practice of ploughing with only two horses instead of teams of six to ten oxen and horses. Before 1754, wrote the Banff minister, the whole country had become a vast common for the grazing of flocks and herds after the harvest had been garnered, and even the best farms had been divided into infield and outfield. But, in 1754, the common pasture on Gallow-hill was feued by the burgh council to the townspeople at ten shillings per acre, with the right to buy the feu duty for 25 years' worth of payment; by 1798 the pasture had become divided and enclosed with dykes, and was letting for £2–3 per acre, and some market gardeners were renting fields at £5–6 an acre.[13] Until about 1790 some 400 acres of land surrounding the burgh of Nairn was cultivated in runrig but then the proprietors gained the right to enclose their portions, and by 1794 some two or three had done that with stone or turf dykes. Several new crops made their appearance around 1754. Two gentlemen farmers, both former army officers, introduced sown grass and turnips to Gartly in Strathbogie in 1770, establishing a field of each, and by the time he came to write the OSA account in 1794 the minister noted how these crops were now on all the farms. The introduction of the potato as a field crop happened in the 1750s, arriving just when a new foodstuff, relatively easy to grow, was beginning to be needed by a population of rural poor who were finding themselves more and more constrained by the change going on around them. By the 1790s it had become a principal component of the diet. An acre of potatoes yielded around 16 bolls, noted the minister of Auldearn, and was the subsistence of the people for one-third of the year.[14] The minister's colleague in Dyke reckoned every boll of tatties planted gave eight bolls to lift. 'About the year 1758, potatoes became a principal article of provision here,' wrote the Revd John Bethune in Dornoch in 1791. 'Now they serve as the chief subsistence of the people during a third part of the year; with many for one half, and with some even for two thirds . . .' The hardiness of the potato and its ability to yield some kind of return on the most unpromising patch of soil was a boon to families hacking a living on the margins of arable ground.

The merging of tenancies and the enclosure of open ground deprived many people of stretches of land they had cultivated or used as grazing. To compensate for this loss, some turned to reclamation of hitherto uncultivated waste. 'Several old people now alive remember the first culture of a space of ground within its precincts that may contain, at present, a tenth part of the whole population,' observed the minister of Kirkmichael parish, an isolated part of Banffshire with a low number of inhabitants.[15] On the Black Isle Mr Mackenzie of Allangrange reclaimed some 80 acres of boggy land for letting as new arable farms. Such experiments did not always repay the effort. The earl of Findlater tackled the 'improvement' of the common land of Green Hill on the east side of the burn of Deskford, and had it enclosed, hedged, ditched, planted with belts of pine and alder, and 'a complete set of farmhouses built' but, after considerable expenditure and repeated efforts, 'his Lordship was discouraged', apparently defeated by the exposure and the soil.[16] Green Hill became pasture and much is now under forest, although Findlater may have approved of that, as he had successfully planted larch in another experiment.

Droving had a long history in Scotland – various pieces of legislation relating to moving herds over vast distances and down into England date

This life-size bronze sculpture by Lucy Poett was erected in 2010 outside the Dingwall livestock mart to commemorate the droving trade.

from the fourteenth century – and a network of drovers' routes covered the country, marking out ways from every corner of the north to the great trysts at Crieff, Falkirk and other centres. The skills associated with cattle-lifting found a legitimate and lucrative outlet in this commerce. The trade underwent expansion during the eighteenth century in response to the burgeoning market for beef in the south of the country. This had its biggest effect in the Highlands but it happened also in the north-east, and it was often the main way in which cottars and crofters in poor areas could earn some cash. Droves from Buchan, Banff and Moray crossed the Mounth to Deeside and then made their way south to Dunkeld, while those from the north-west usually came together at Muir of Ord before proceeding by way of Strathglass or Strathspey. The minister of Wick noted in 1793 that 'the county at large, as well as this parish in particular, abounds with black cattle; considerable numbers of which young and old are purchased by drovers at from 40s to 50s per head, and are driven to Falkirk, Edinburgh and England', a remark that could stand for most of the parishes in the north.[17]

The formation of new villages became almost a standard item in the package of improvement measures implemented by the enterprising lairds. The process was similar everywhere: a surveyor decided the boundaries of the lots, a factor or a lawyer, or probably both, drew up regulations for the tenants, the lots were assigned, perhaps by lottery, and the village stumbled into existence, echoing in its birth the formation of the royal burghs centuries before. According to the Revd Alexander Humphry, the assistant minister of Keith in the 1790s, Lord Findlater 'divided a barren muir' in 1750 and feued it out in small lots on a regular plan, creating what had become 'a large, regular, and tolerably thriving village called New Keith, containing 1,075 inhabitants'.[18] The 'tolerably' hints at growth slower than perhaps hoped for, but growth nevertheless. 'This village is the residence of all manufacturers of note in the parish,' continued Mr Humphry, adding 'according to the success of their business, therefore, it must either prosper or decline.' This was pretty well the story of all the planned settlements. Established in 1764, Strichen had 200 inhabitants – weavers and other tradesmen – in 1793. Grantown-on-Spey was only 20 years old in 1793 and already had close on 400 residents on what had been 'a poor rugged piece of heath'. In the shadow of the new settlements some of the older villages faded – the original kirktown of Keith had become 'almost a ruin' by 1790 – but others continued and grew. Huntly 'has surprisingly increased within these last 50 years in population and industry,' wrote the minister in 1794.[19]

Textile manufacture was a favourite choice for economic activity in the new settlements. In 1748 the earl of Findlater brought two or three 'gentlemen's sons' from Edinburgh who knew the linen business, and gave them each £600 interest-free for seven years, along with work premises and looms; this local linen industry was still flourishing in the 1790s. The proprietor of Cromarty, George Ross, started a hemp factory in the town in 1773. In 1752, Findlater started a linen and damask manufactory in Cullen. In 1794 burgeoning Huntly had 52 employed in flax dressing, 209 weavers and a small cotton factory. Johnstons of Elgin began as a cashmere business in 1797 at Newmill on the Lossie. Some of the textile production may have been only a step above traditional cottage output, such as the weaving of plaiden and 'coarse tartan, with a kind of broad cloth and duffle' in Ardclach,[20] but the scale was less important than the income it could bring to a family. In the end the local ventures were often brought to an end by the availability of cheaper cloth from outside the region: cheap cotton, flax from the Netherlands and Irish flax coming into Glasgow brought about a decline in the dressing and spinning of flax in Keith. In Fraserburgh imported Dutch flax was used in the manufacture of linen yarn, but in Banff thread-making from Dutch flax was abandoned by 1798 in favour of the manufacture of stockings, which was giving work to as many as 560 people. Banff could also boast a soap and candle factory, brewing and rope and sail manufacture. Like all the coastal towns and villages the sea on its doorstep brought better opportunities of every kind, in an age when maritime transport had greater significance than it does now. And of course there was fishing.

The gentlemen improvers did not neglect the resources of the sea. The relatively shallow Firth – the average depth of the predominantly sandy bed is 54 metres, the deepest spot 213 – was a fruitful fishing ground. In 1721 Alexander Hepburn listed the species caught off the Buchan coast: 'Killing [large cod], Leing, Codfish small and great, Turbet, Scate, Mackrel, Haddocks, Whittings, Flooks, Sea dogs and Sea Catts, herrings, seaths, podlers, Gaudnes, Lobsters, partens and several others'.[21] Alexander Garden of Troup, who was to bestow his name on the village of Gardenstown in 1720, made a list in 1683 but extended it to include sea fowl – scath, badoch, coot, sea coulter, taster, maw, kitwiack and whap – and more shellfish –

The Deveron estuary separates Macduff in the foreground from Banff to the west.

buckie, mussel, clam and cockle.[22] In the eighteenth century sea birds were still occasionally taken as food, the kittiwake 'whilst young . . . no better flesh eaten'. In Caithness, a favourite for eating was the lyre, as the shearwater was called.

In the 1650s Robert Gordon of Straloch had hard words to say about the fishermen: 'men from the dregs of the populace who have given themselves up to this life follow the fishing for daily requirements and not for gain from trade,' he wrote.[23] Perhaps the coast had become a refuge for outcasts in the troubled mid century or perhaps fishermen were an independent-minded lot whom landowners found tricky to handle but Gordon's observation sits oddly with the fact that in later times fishing communities were to display a morality and adherence to religion enough to satisfy the most ardent Presbyterian. Perhaps Gordon's real purpose was to spur greater use of the sea, for he goes on to lament the foreigners, especially the Dutch, who 'make great profit every day before our eyes from the capture of herring and other fish'. The standard method of catching white fish was line fishing. It is a relatively simple technique – long lines with several hundred hooks spaced at fathom intervals are trailed in the sea – but it is time-consuming and labour intensive. As the gathering of baskets of the mussels or other creatures used as bait meant long hours on the shore in all weathers, the line fishing was essentially a family business, with the men, wives and bairns all doing their share. It was not work that a person from outside could easily adapt to, and this led to the fishing communities evolving as tightly knit societies distinct from their landward neighbours. Along the southern shore of the Firth, it was common to find in any fishing village only a few surnames but an extensive list of tee-names or by-names, a mark of belonging that could carry through the generations.

The economic possibilities of the fishing were not lost on a few lairds and several set about establishing new fishing settlements, although fishing families had long shown a propensity for coastwise migration possibly on their own initiative. Portknockie was founded in about 1677 by families from Cullen. The first houses in Buckie were built in 1645. Fishers from Fraserburgh established themselves at Findochty in 1716, and the laird of Rannes settled five families at Portessie in 1727. According to the minister's account in the OSA, there was only one fishing boat at Avoch in around 1700 but in the century since then a sizeable seatown had grown. It is highly likely that the fishing people in Avoch moved there from elsewhere on the Moray coast; their Scots dialect, distinctive in a county where

Highland English was the norm, gave rise to several colourful stories about their origins. Likewise in Nairn, the fishertoun and its inhabitants remained distinct from the rest of the town for a long time. The growth of the fisheries and the traffic of trading vessels were encouraged by the building of harbours. The harbour at Pitsligo dates from about 1679, the old harbour at Portsoy from around 1692 – although there was an older haven for a century before that, and Cullen harbour existed in 1690. The Fraser lairds of Philorth expanded the village of Faithlie from 1546 onwards to form the harbour and town of Fraserburgh. Banff harbour dates from the 1770s, Lossiemouth from 1703 when a pier was completed at the river mouth. On the north side of the Firth, Cromarty gained its harbour in 1785, Brora had a 'tolerable' harbour in 1793, yet Wick lacked a decent haven and the Caithness grain trade operated mainly through the nearby village of Staxigoe. At many places, boats were simply beached and this remained the case until the expansion of harbour construction in the nineteenth century.

Cod, ling and other white fish could be salted and dried in the open air to produce hard, long-lasting stockfish. Oil from fish livers was also a valuable commodity. At the end of June cargoes of dried fish would be

Part of the fishing village of Gardenstown. The houses under the brae are typical of the Firth coast in being built with their gable ends to the sea.

155

taken from the Buckie shore to market in Fife and the Firth of Forth, producing in the early 1790s for the fishermen an income of £8 to £12 per man, with half-pay for the boys, wages considerably higher than the £6 to £8 a farm servant could expect to earn. In the early autumn the Buckie fishermen crossed the Firth to Caithness to join in the herring fishery for six weeks, where they could enrol in the government bounty system or engage with a curer on another basis – perhaps ten shillings a barrel of fish and a bottle of whisky per day. The lack of a suitable harbour on the Buckie side – the fishermen dragged their boats up the beach – limited herring fishing closer to home. There was also a lucrative lobster fishery with the catches being sold to English merchants such as Messrs Selby of London or the Northumberland Fishery Society. The Rathven minister spoke of the 'amazing quantity of lobsters' to be had on the Caithness shore, but he also recorded that 7,913 lobsters had been landed at Portknockie in 1792 for the Northumberland company. The Duffus minister recorded a barely credible 60,000 lobsters caught for the London market last season in his parish. A special note should be made here of the importance of salmon, caught in the sea or estuaries in stake nets or sweep nets, so lucrative that this became the one fishery where lairds had carefully protected legal rights. About 130 men were employed in the salmon fishery at Bellie on the east side of the Spey in a business rented from the duke of Gordon for £1,500 per year. The salmon were exported to London on ice – stored in the large ice-house – or steeped in vinegar in special barrels called kits. What could be termed the traditional fisheries in the Firth were poised to be overshadowed by the tremendous expansion of herring fishing in the nineteenth century but that story belongs to the next chapter.

Boatbuilding took place at almost every settlement around the coast, a steady production of seagoing vessels, for fishing as well as for trade, by local craftsmen. A more substantial shipbuilding industry grew up in Kingston and Garmouth at the mouth of the Spey, using the Scots pine timber floated down the river from the forests of Glenmore and Rothiemurchus and, later, timber imported from the Baltic. Although there had been a relatively short-lived venture by the York Building Company to log in Strathspey, the industry did not achieve its maximum size until two Hull timber merchants, Dodsworth and Osbourne, bought the Glenmore forest from the duke of Gordon in 1786. The minister of Speymouth, James Gillan, described their operation for the OSA, the floating of the logs down the river in rafts at 30s per raft by gangs of men, mostly from Ballindalloch: 'These men have 1s 2d

a day, besides whisky, and there will sometimes be from 50 to 80 employed at once in the floating'.[24] Speymouth shipbuilding reached its zenith at the same time as the herring fishery in the nineteenth century.

The mention of whisky – it frequently features in the parish accounts of the 1790s, usually to be deplored by the clerical authors such as the Revd William Sutherland in Wick who thought 'There is enough of malt and too much whisky, which is prejudicial to the morals and constitutions of many'[25] – is a reminder of how ubiquitous and important it was. There were 20 licensed stills in the county of Moray alone, or 19 and a brewery, depending on one's source.[26] This does not take account of illicit distilling or smuggling, by its nature beyond the reach of statistical observation. In the sixteenth century, with the brewing of ale, it was a widespread cottage industry, often the resort of older people as a less strenuous way to make a livelihood. In 1564 the burgh court in Inverness licensed four makers of spirits, in 1567 eight, a much smaller number than the 70-odd licenses

The fishing village of Portknockie dates from 1677 but the Picts had a base here in the seventh and eighth centuries, sited on the rocky promontory of Green Castle in the lower left of the picture.

This impression of Cromarty, made in 1801, exaggerates the proximity and height of South Sutor but includes the long building of the hemp factory near the harbour.

granted to ale brewing in recognition of the dangerous strength of the product. In the cold and troubled seventeenth century, Sir Robert Gordon of Straloch noted how people of every rank drank spirits in large quantities: 'Hardly anyone of better quality abstains, nor do the ladies escape this disgrace. Men travelling in the stormiest time of winter . . . fortified with a flask of this liquor . . . perform the longest journeys on foot without any inconvenience'.[27] In 1690, Duncan Forbes of Culloden was given the right to sell free from customs duty the whisky distilled on his lands of Ferintosh on the Black Isle, as compensation for damage wrought on his property by Jacobites. This lucrative and unique privilege angered every other distiller in the country and it was eventually revoked in 1786, with compensation of £20,000 for the Forbes family. Where there were stands of birch there was sometimes a custom of making wine from the sap, a liquor called *fion na uisg a bheatha* by the Kirkmichael minister in Banffshire, a reminder that the phrase 'water of life' had a punning resonance in Gaelic with the word for birch. The Kirkmichael minister revealed himself to have an unusual interest in folk traditions and he also talks about the lost recipe for heather ale.

The search for new economic projects in the age of Improvement also encompassed minerals. The York Building Company opened an iron ore mine in the Banffshire hills in 1736 but abandoned it after three years. The stone of Portsoy had enjoyed an export to France for a time, according to the OSA account, before it became unfashionable and a shipload of it was left neglected on the Seine. The lack of major iron ore deposits and coal, however, kept large-scale industry out of the Moray Firth region and what minerals were present in quantity became valuable for agricultural fertiliser and building stone rather than for metal extraction, ensuring the land remained predominantly devoted to primary production.

Along with the changes in agriculture and commerce came improvements in transport and what today is termed infrastructure. In the wake of the 1715 rising, General George Wade and his successor William Caulfeild drove new roads through the mountains. Theirs was the first major initiative to break away from the old route that came to Moray by way of Buchan and in time would alter the geographical orientation of the Firth communities. Wade in effect laid the foundation for the modern A9, although his route over the Corrieyairack Pass from Strathspey to Fort Augustus, itself a considerable achievement, proved less significant in the long run. Caulfeild's route from Deeside to Fort George via the Lecht and Grantown, now mostly the A939, is still a spectacular journey. The military roads fell into disrepair towards the end of the eighteenth century, by which time other efforts to improve transport were being made. The large rivers posed serious obstacles to land travel. There was no bridge over the Spey at Fochabers in 1793 when the duke of Gordon started a subscription fund to have one erected at Boat o' Bog, raising some £3,000 within a year before war broke out with France and the plan was shelved. The ferry at Boat o' Bog, now the site of the A96 bridge, charged a halfpenny for a single traveller, twopence for a man with a horse and 2s 6d for a chaise and pair – the prices rose in time of flood, presumably to reflect the dangers of the crossing. In 1795 another proposal existed to throw a bridge over the Spey further upstream at Boat o' Brig. The Spey was bridged at last in 1801–04 by an elegant structure designed by Thomas Telford. General Grant in 1792 had a bridge built over the Avon about half a mile above Ballindalloch. The Findhorn river was 'rapid and frequently impassable everywhere excepting at Dulcy bridge,' noted the minister in Ardclach in the OSA, adding that although there were ferries in use the boatmen were not as skilful as could have been desired, and 'many lives are lost'; 23 had been drowned in this way since his arrival in the parish.[28]

In his description of Forres in 1796, the minister recorded a startling fact and one that still raises a smile – the number of tea kettles in the town

An impression of
Nairn made in 1804.

Opposite: This map
of the Black Isle,
reproduced from
Sir John Sinclair of
Ulbster's general
view of agriculture in
the northern counties
published in 1795,
shows a landscape in
transition. Enclosed
fields and plantations
are appearing
around the houses of
the lairds but a large
tract of common land
still stretches over
the high ground of
Mulbuie. The map
incidentally shows
how the peninsula
was divided over
three sheriffdoms:
Ross and Cromarty,
and the enclave of
Ferintosh that came
under Nairnshire.

had risen from three to 300 in about 50 years. There is more information in this vein – then there had been only six people with hats instead of the traditional blue bonnet, now there were over 400 folk with new headgear. Thirty years before, a suit of holiday clothes for a servant had cost 30 shillings, now the price was £5. In the neighbouring parish of Dyke, there had been an 'alarming progress' in the use of tea.[29] All around the Firth one could find similar instances of the rising standards of living. Elgin could boast 44 shops selling imported goods. Wages had gone up. A harvest servant, a man hired for a day's labour in bringing in the crop, could expect 4d a day with victuals in 1750 in Forres; in 1796 the going rate was 10d with two meals. A journeyman mason could now expect 20d a day, almost double what it had been in mid century, and a woman's wage had risen from between 8s 4d and 10s to 18–20s per six months. The price of labour had been 'amazingly raised' in the last 30 years in Elgin – a ploughman, the elite among farm labourers, used to get 40–50s a year, now it was £5–7.[30] Rural wages lagged behind those in the towns but the upward movement of urban incomes was dragging farm wages in the same direction, and that in turn was diverting people away from the old ways. The Revd John Grant in his manse in Kirkmichael in the Banffshire hills noted that his parishioners had formerly neglected industry and agriculture in favour of traditional cattle-raiding and other such pursuits but were now affected

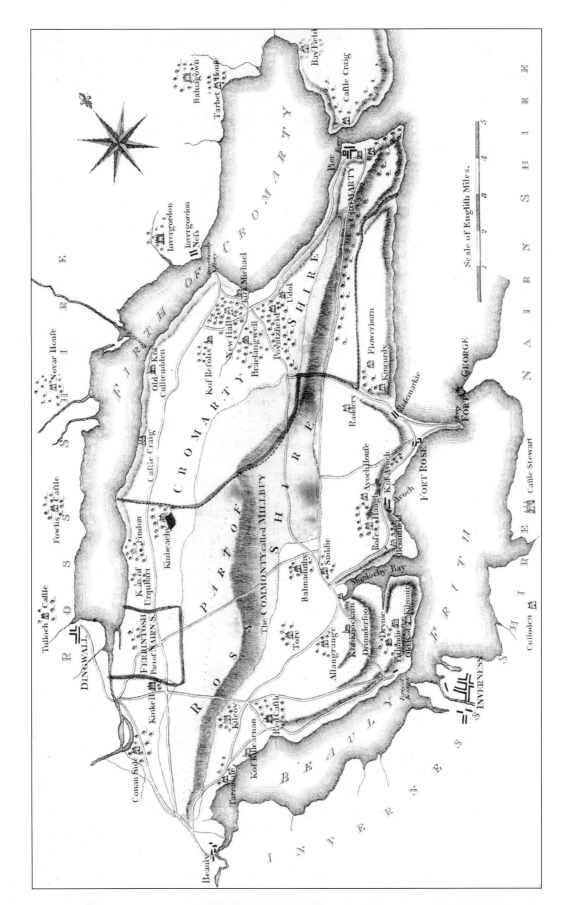

by the 'the spirit of commerce . . . with its natural concomitants, avarice and selfishness'.[31] Mr Grant had quite a bit to say on this theme – enough to make one wonder how his sermons went over. Several of his clerical colleagues showed nostalgia for a simpler past but most, while decrying the growing materialism, gave a cautious welcome to the improvement in the living standards of their flocks. 'Formerly their hair flowed in easy ringlets over their shoulders,' wrote Mr Grant of the women, 'not many years ago it was bound behind into a cue, now it spreads into a protuberance on the forehead . . . sometimes, it is plain and split in the middle. But who can describe the caprice of female ornament more various than the changes of the moon!'

Although schools were widespread in the late 1700s, the Knoxian ideal of one in every parish, although reinforced by statute in 1696, had not been completely realised in the north. Sutherland lacked facilities in the 1790s. In Brora in 1793 there was a parish schoolmaster but no parish school – the master 'teaches children reading, writing and a little arithmetic in his own house'.[32] In neighbouring Loth, to the north, a newly built schoolhouse had put an end to the difficulties in obtaining a master, and now 30 to 40 scholars were attending. The system called for the heritors in each parish to contribute towards the salary of the schoolmaster but he could also have other sources of income. In Latheron parish, for example, the master received 100 merks per year for teaching – the standard rate – but also had £20 Scots for officiating as session clerk and precentor, with an additional 6d for every baptism, 6d for each certificate and 1s 7d for each marriage. Latheron parish stretches for 27 miles along the south-east Caithness coast and is divided naturally into a series of straths. Travelling to one central parish school would have been impossible for pupils – this was a problem common to many rural parishes throughout the north – and there were, in Latheron, three or four other schools 'supported by the inhabitants' in the distant parts, as well as two schools run by the Society in Scotland for the Propagation of Christian Knowledge. The latter, founded in Edinburgh in the early 1700s to further education in the Highlands, supported schools mainly in rural areas but also managed a few town-based establishments such as Raining's School in Inverness. Private schools were common in the towns. Wick had 'several private teachers' and 'no less than five charity schools'.[33] The towns were also where advanced institutions of learning were founded through philanthropic effort – Inverness, for example, began to raise funds in 1787 for what would become the Royal Academy. Most children were probably receiving some basic formal education by the late 1700s, whether in the schoolroom or through a mixture of lay tuition and religious instruction.

Many of the OSA accounts record a major improvement in health through inoculation against smallpox. Dr Edward Jenner published in 1798 the results of his experiments on using cowpox as a vaccine against the more virulent smallpox but by this time a fair proportion of the Scottish population had already been inoculated. The basic technique was simple, and involved scratching the skin and inserting in the wound a tiny amount of matter prepared from a scab from an infected individual. The risk of catching a fatal dose of smallpox from the crude inoculum was of course high but the fear of the disease, likely to flare up every few years in an epidemic, was usually enough to overcome any doubts. A bad outbreak in Dingwall in 1789 overcame prejudice and opened 'the eyes of the lower classes' to the boon of inoculation'.[34] 'Inoculation universally takes place with great success,' wrote the minister of Cromdale.[35] In Mortlach in Banffshire, the minister lamented the dislike the people had for inoculation, 'the neglect of which, though it is in use rather more than it was, makes this very infectious and virulent disease [smallpox] frequently mortal'.[36] In a smallpox epidemic in Thurso in 1796, 25 per cent of the victims died; John Williamson, surgeon to the local fencible battalion, carried out a plan of general inoculation but had to call on the clergy to help dispel the belief that the treatment was a 'provocation to Divine providence'. With a clerical seal of approval, Williamson inoculated up to 120 people in a single day. In Inverness a square wright [carpenter] was inoculating the lower classes at no charge and perceptively was advising fresh air, a simple diet and cleanliness as measures to ensure health. Changes in diet and in public health, with the survival of more children into adulthood, encouraged an increase in population evident across the whole country as the century drew to a close.

There are no reliable statistics for the fall in the use of Gaelic but circumstantial evidence tells of a language in retreat. By the end of the century in Buchan, the universal tongue was Scots, and Banffshire was also shifting that way. In Boharm, where the Fiddich joins the Spey, 'the Scotch' was the only language spoken. In Botriphnie people had forgotten when Gaelic had been spoken 'in this part of the country'.[37] In Mortlach, 'there is hardly a word of Erse now spoken in any part of the parish', except possibly in the Glenrinnes area, admitted the minister, where 'the inhabitants do

The walls of Old St Peter's, rebuilt in the 1700s on a medieval base, in the village of Duffus are surrounded by a crowded graveyard. The slim parish cross is silhouetted against the yew. The grounds contain an 1830 watch-house where men could guard against the depredations of grave-robbers.

also most retain the look, manners and genius of the Highland Caledonian, as appears from their dress, their vivacity, their social and merry meetings, their warm attachments, their keen resentments, their activity on occasions and indolence on the whole, their intelligence and their love of their country'.[38] These words written by the Revd George Gordon, who had already transferred to Aberdeen by the time his account was published, smack of cultural frontier, and it is hard to decide whether the minister liked the people or was wary of them. In parts of Moray and Nairnshire, Gaelic was to be found as a daily tongue only in the upland parishes. Moving west, in Cawdor, 'Erse and English are equally spoken'[39] and in the Inverness-shire parish of Moy and Dalarossie, Gaelic only.[40]

All through the seventeenth century at least one of the ministers in Inverness had to be fluent in Gaelic. During the reign of Queen Anne, which ended in 1714, Gaelic was spoken by about three-quarters of the population of the town and Inverness remained a bilingual community, as fitted its self-styling as the capital of the Highlands. The kirk session decided in 1791 that it would be better to build a new Gaelic kirk than continue to repair the existing one, and settled on a design big enough to accommodate

1,200 people, only to enlarge it a few months later to take another 200. Among the merchants, lawyers and their ilk, another language transition took place – Scots declined in favour of a more southern English. This can be seen in the minutes of the burgh council and other contemporary documents. What were in the 1500s called the Kirkgate, the Briggate and Damisdail [Doomsdale] became Church Street, Bridge Street and Castle Street. Exactly what powered this linguistic shift is unclear: the presence of English Cromwellian troops for a decade may have contributed but on its own would not have driven the change. For the origins of Invernessian English we have to look rather to the effects of a few generations of school and Bible education on a population whose mother tongue tended to be Gaelic. Inverness, unlike Moray and northern Caithness, did not have time to develop a distinctive Scots before English came in.

Moving north of Inverness, the language map at first shows Gaelic spoken everywhere but in a few specific coastal locations, exceptions which appear to have been of long standing. There was a pocket of Scots speakers in Avoch in the fishing community. Cromarty was Scots-speaking, 'this being one of the three parishes in the counties of Ross and Cromarty in which, till of late years, the Gaelic language, which is the universal language in the adjacent parishes, was scarce ever spoken', but it acquired a sizeable Gaelic-speaking community and a Gaelic chapel in the late 1700s through an influx of workers.[41] In Alness the Revd Angus Bethune said 'Gaelic . . . is generally spoken by the country people' and although the 'heritors and higher ranks' seldom spoke it some of them understood it enough to converse with their tenants.[42] Bethune noted that English had made considerable progress in the preceding 20 years because of the three schools, the parochial one near the kirk and the two SSPCK schools in the uplands. Dingwall had only the parochial school; Gaelic was still the language of the common people but most were bilingual, as was the general population around the coast of Ross as far as Strath Oykel. Gaelic was prevalent along the Sutherland coast and the south-eastern part of Caithness. Curiously the Revd Robert Gun, the author of the description of Latheron in the OSA, makes no mention of Gaelic but we know from the account of Wick by his northern neighbour that Latheron was one of four Caithness parishes – the other three were Halkirk, Reay and Thurso – where Gaelic was used for divine service and therefore was undoubtedly the daily tongue of most of the ordinary inhabitants.

One question that suggests itself for the 1790s is whether the shifts

This stone monument beside the A9 at Loth commemorates the killing of the supposed last wolf in Sutherland in around 1700. Wolves appear to have survived a little longer in the central Highlands, and one tradition has it that the last wolf in Scotland was shot in the Findhorn valley in 1743.

in education, language and economy were fuelling changes in political attitudes. An extension of this question – whether or not religious views began to change – would seem to be answered in the negative, at least for the time being. Abercromby Gordon, the minister of Banff, made the comment that the people had not been unsettled by Paine's *Age of Reason*. The writings of Thomas Paine, the Norfolk-born pamphleteer and radical thinker, were clearly in wide circulation. Paine had published *Common Sense* in 1776 in defence of the American Revolution, and in 1791 *Rights of Man* about the French Revolution. *The Age of Reason*, issued in 1793, was an attack on established religion and a plea for rational free thinking; hence Gordon's singling out of this particular work. There appears to have been no large-scale opposition on the southern side of the Firth to the throwing together of farms and the enforced flitting of former tenants but the feeling existed among the better educated that one day this tolerable state of affairs might end. Elizabeth Grant described how her father in Rothiemurchus decided that some knowledge of the law was 'necessary to the usefulness of a country gentleman . . . the French Revolution, in the startling shake it had given to the aristocracy of all Europe . . . had made it a fashion for all men to provide themselves with some means of earning a future livelihood, should the torrent of democracy reach to other lands'.[43] The torrent was yet but a few trickles in the Firth lowlands but these were increasing in strength. Copies of Paine's provocative writings were in circulation in southern Scotland in various forms and inevitably were filtering north. Apart from the normal channels of commerce, there were now large numbers of men travelling as soldiers or seamen and bringing back ideas. In 1793 copies of *Rights of Man* were found in Stornoway at the time of rioting against Mackenzie of Seaforth's attempts to recruit for his fencible corps. Judging, however, by the writings of the parish ministers at this time, while admitting they had every incentive to praise the quiet conservatism of their flocks, radical protest was rare and the customary hierarchies prevailed. What Hugh Miller had to say about his fellows in Cromarty probably could be fairly applied to the great majority around the shores of the Firth. 'The people . . . who lived ninety years ago . . .' – Miller was writing in the early 1830s – '. . . were quite as unskilled as their neighbours, and thought as little for themselves. They were but the wheels and pinions of an immense engine; and regarding their governors as men sent into the world to rule – themselves, as men born to obey . . .'[44] Then came the political tumults across the Atlantic and the Channel. 'The crash of a falling throne awakened opinion all over Europe,' wrote Miller. 'The young inquired whether men are not born equal; the old shook their heads and asked what was to come next?' Newspapers were beginning to circulate, says Miller, adding that a crop of young Whigs began to shoot up 'like nettles in spring'.[45] An ironic dismissal in that Miller himself, although politically a conservative, was to play a leading role in social change only some ten years after he expressed these opinions.

We see during the latter years of the eighteenth century an old society opening itself up to new ideas and the old hierarchies starting to weaken. For all the progress, there was still risk of famine. On the south side of the Firth, the changes had bedded in but, as the calendar turned towards 1800, the northern side had still to experience the full impact of what the parish minister of Canisbay in Caithness in 1791 hailed 'the advancement of civilisation'.[46]

# CHAPTER 11
# SHEEP, ROADS AND HERRING

In 1791 agricultural change had only begun to touch the parish of Dornoch. 'There are no field turnips cultivated here,' noted the minister, 'and scarce any sown grass raised, except in two or three places, the seats of men of property; which are also the only inclosures in the parish . . . the arable ground is, for the greater part, in constant tillage'.[1] In the neighbouring parish of Golspie, most of the large farms were enclosed and engaged in mixed farming, growing the full range of crops and raising black cattle on the coastal pasture and sheep on the hill-ground common. Beyond Golspie, the band of arable land along the coast shrinks steadily in width until it almost disappears and the hills crowd towards the sea, shouldering into the Ord, the high moorland forming the boundary between Sutherland and Caithness. Here the acres available for improvement through the enclosure methods developed in southern counties were very limited but there was an alternative model for the landlord to adopt to boost his rental income, one already being rolled out, large-scale sheep farming. *Am Bliadhna nan Caorach* – the Year of the Sheep – is a phrase that rings across two centuries. For many it encapsulates a major trauma in Highland history, the Clearances, when according to the popular view the ordinary folk of the clans were ousted from their ancestral lands by cruel landlords to make way for sheep farming. The actual year in the phrase is 1792. In that summer small tenants and subtenants in the uplands in Easter Ross and around Strath Oykel, aggrieved over the way they were losing their grazing, resolved to band together to drive imported flocks in the care of southern shepherds out of the country. The authorities and the gentry panicked, fearing the start of another clan uprising or revolution on the French model, and called out the Black Watch from Fort George to restore order. Five men were arrested as ringleaders, hauled before the circuit court in Inverness, found guilty of disturbing the public peace and damaging property, and given harsh sentences, but some person or persons unknown helped them to escape from the Inverness Tolbooth in dead of night and disappear back to their own people.

So much has been written about the Clearances, especially in Ross and Sutherland, that the events themselves have almost become smothered in later opinion. It is impossible now to establish with precision how many families and individuals were driven from their townships to make room for sheep but a listing of some evictions in the parishes touching on the Firth indicates the extent of the upheaval: Rosehall in 1788; Easter Ross in 1792; Kildonan and Clyne in 1813; Strathnaver in 1818–20; Culrain in 1820; Dunbeath in 1835; Glencalvie in 1845; Strathconon in 1840–50; and Greenyards in 1854.[2] Many episodes of clearance attracted little attention either because they were peaceably done or because they happened as a slow, piecemeal erosion of the existing communities. When he came to write about his parish – Alness – for the *Statistical Account of Scotland* (OSA), the Revd Angus Bethune struggled to give an even-handed account of the trouble in *Am Bliadhna nan Caorach*. Sheep farming had been lately introduced on the higher ground, he wrote,

> but the plan, however justifiable in itself, and on the part of the proprietors, was unpopular . . . and excited a disorderly and tumultuous spirit among the country people which it became necessary by legal and forcible means to suppress. This gave rise and rapid circulation to a report, as injurious as it was groundless, that the proprietors treated these poor tenants with oppression and cruelty. In justice, however, to the proprietors of *this* [sic] parish . . . it is proper to assure the public that such tenants as had been removed . . . were otherwise provided in farms by these gentlemen either on their own estates or on some others . . .

Bethune's emphasis on the 'this' in reference to his parish reveals his dilemma – trying not to appear generally pro-landlord but at the same time desiring to be supportive of the principal heritor in the parish, General Sir Hector Munro of Novar, who, in the minister's view, had carried through wonderful improvements to the family seat and had not increased any of the tenants' rents.[3]

The man for whom Bethune wrote his description of Alness, the originator and editor of the OSA, was Sir John Sinclair. He had been born in Thurso into the Ulbster branch of the Sinclairs in 1754, traditionally the same year in which the potato was introduced to his native Caithness. After studying law, Sir John launched on a career in politics, becoming MP for Caithness in 1780 in the first of several seats he was to hold in a long, active, public life. He is remembered now for his passionate enthusiasm for a whole range of subjects relating to economics and development, an enthusiasm he was tirelessly eager to share with others. He was a member of a slew of scientific and arts societies, he founded the British Wool Society in 1791 and became first president of the Board of Agriculture, he laid out plans for the town of Thurso and other settlements on his estates in Caithness, he introduced to the English language the word statistics – by which he meant 'inquiry into the state of a country, for the purpose of ascertaining the quantum of happiness enjoyed by its inhabitants, and the means of its future improvement'[4] – and conceived the idea of compiling a statistical account of all Scotland, and he brought to the Highlands the Cheviot sheep.

There had always been sheep in the north, small animals similar to the Soay and Shetland breeds, held in lesser regard than the cattle in the pastoral upland culture. The breed that troubled the Easter Ross tenants in 1792 was the Linton or Blackface. The 500 Cheviots Sir John brought from the Borders to his wife's estate at Langwell near the Caithness–Sutherland border represented another innovation, a large breed with better quality wool but not so hardy as their predecessors. Hardy enough, however, for the Firth coast, where they flourished, surprising both their shepherds and the locals, who thought the sheep would need to be housed in winter. The introduction of the Blackface and the Cheviot – and the buoyant market for wool – confirmed large-scale sheep farming as the most attractive option for the improving landlord. Sir John believed that the existing tenantry, seeing the value of the new livestock, could also be encouraged gradually to switch from cattle rearing to the raising of sheep, while some of the rural tenantry moved to his new planned villages. To him the eviction of large numbers of the rural people to make room for incoming shepherds and their flocks was a barbaric way to go about things but in practice even he had to resettle tenants from Langwell to an exposed stretch of cliff top at Badbea to tiny crofts where they laboured for a few decades before emigrating. Sir John Lockhart Ross of Balnagown entertained the ambition to build a village and mills in Kincardine where processing the wool from the Linton flocks would provide new employment for the tenants but that idea also fell by the wayside.

A few proponents of improvement were able to remind their fellows that the interests of the peasantry need not be subordinated to the new ways. Sir James Grant, the laird of Glen Urquhart, helped his tenants improve their own small stocks of sheep by buying tups from western sheep farms, gave long leases to some tenants, enclosed their land, and provided rye grass and clover seed and access to a limestone quarry. He had a lint mill built and made efforts to establish linen production. George Dempster of Skibo also tried to involve his tenants in improvement. He came from a wealthy family from Dunnichen near Dundee and in his youth was a member of the Poker Club and the Select Society, two of the Edinburgh clubs where strong drink and intellectual conversation flowed in equal measure in the late 1700s. After a spell as MP for the Fife and Forfar burghs and a public career in London that included a directorship of the East India Company – he had to resign for pushing the idea that it should relinquish imperialist rule in favour of disinterested mercantilism – and being a founder director of the British Fisheries Society. In 1786 he bought the rundown estate of Skibo and put into action his own notions of improvement. He found his newly acquired tenants to be living in miserable circumstances: 'An iron pot for boiling their food constitutes their principal'.[5] Dempster also managed the neighbouring estate of Pulrossie that had been bought by his brother, an officer in the army. At the request of Sir John Sinclair, he wrote about his plans in an appendix to the account of the parish of Criech in the OSA. The 18,000 acres of the Dempster lands stretched along the hilly northern shore of the Dornoch Firth west from Ard na Cailc, where the modern bridge spans the grey water. Dempster thought that around 200 families, apart from those on the mains farms, called it home. These tenants cultivated small patches of oats and potatoes and raised cattle to pay their rents, keeping them indoors in winter and feeding them on straw. The tenants' houses were built for the most part from turf, cut from the ground in 'the form of large bricks'. Every three years the earthern part of the building was thrown on the midden and new walls erected in its place. Young men and women were in the habit of migrating south in the spring in search of work, coming home before winter, while those who stayed at home tended to the family fields and livestock, and cut the peats. It was a poor, hard-scrabble existence, in Dempster's view, and he stated that he had no intention of increasing rents before he had encouraged his tenants to improve their plots and acquire better houses. A key feature of his project was the establishment of a cotton spinning factory on the Firth and, to this end, he raised £3,000 of capital from among his Glaswegian acquaintances, planned to create two villages, and brought in instructors to teach the locals how to work the spinning jennies and looms. The future of the estates also included secure tenancies, all rents to be paid in cash, forestry plantations on waste ground, – all contributing to what Dempster called – in capitals for emphasis – the HAPPINESS OF PEOPLE PLACED BY PROVIDENCE UNDER OUR CARE. There was no need, he continued, to drive out the people to make room for sheep. In the event, the cotton mill at Spinningdale operated for a few years but, after a fire in 1809, it closed its doors. George Dempster deserved more success. Some of his ideas previsaged some of the terms of the first Crofting Act in 1886, and he continually tried to convince other landowners of the rightness of his concepts before his death in 1818 at the age of 84.

The most notorious clearances – those that produced the now stereotyped image of the burning crofts of the evicted – took place in Sutherland. The statue of the duke of Sutherland on Ben Bhraggie and now the Gerald Laing sculpture of an evicted family at Helmsdale are tangible reminders of this unhappy period in the county's history. The first thoughts about introducing sheep to improve local agricultural output can be dated to 1738 and there were sheep 'of the English breed' at Dunrobin in 1746.[6] Sheep farming on a large scale was mooted for the pasture lands of Lairg and Shinness in the 1760s but the 'tutors', the men who managed the estates on behalf of the young countess Elizabeth after the sixteenth earl died in 1750, put the idea aside. It was not until the countess married the English millionaire the marquis of Stafford, later the first duke of Sutherland, that major schemes to improve the lands came into play. These involved shifting the tenantry from their inland townships to the coast where they could find new, more rewarding employment in a growing fishing industry, at the same time freeing broad acres for profitable wool production. The first implementation of this programme took place in 1807 at Lairg. The evicted families – some 300 people – were first offered places on the shore of Loch Naver and later on the north coast but many refused to accept these plots and emigrated. This discontent should have been a warning to the landlord but instead of ameliorating or abandoning the general policy the countess replaced her factor with two men whose names would become forever linked with oppression in the north – William Young and Patrick Sellar.

Both came from Morayshire, both were thoroughly in favour of agricultural improvement and both had little sympathy with the Gaelic-speaking tenants on the Sutherland estates. Sellar, born in 1780, studied law in Edinburgh and joined his father in his solicitor's practice in Elgin where he rose to be procurator fiscal. William Young, also from Elgin, was described by Joseph Mitchell, the roads engineer, as 'a man of clear head, very enterprising and energetic'.[7] He had used the proceeds from a successful corn business to buy Inverugie estate for £13,000. Part of the land had been rendered barren by blown sand but Young devised a method for trenching up the buried arable soil and restored the estate to fertility. He also built the harbour and village at Hopeman in 1805, and finally sold the estate for £31,500. Young and Sellar crossed the Firth to Dunrobin and, in 1809, took over the rental of the 300-acre farm of Culmailly near Golspie, acquiring it by offering a considerably larger payment for the lease than their rival bidder, Robert MacKid, the sheriff-substitute of Sutherland, the man who was later to lead the prosecution against Sellar in his famous trial. The newcomers drew up far-reaching plans for their farm that included drainage, enclosure, threshing, flax and wool mills, and

People evicted from Glencalvie in 1845 took shelter in the graveyard around the church at Croick. Messages scratched on the glass of the church windows are a reminder of their sojourn amid the tombstones of their forebears before they left the area forever.

173

villages with manufactures. Not all of this came to pass but the obvious drive of the two men impressed the marquis of Stafford and his wife to the extent that Young was appointed their new factor with Sellar as his assistant. The incidents that have ensured the retention of Sellar's name in the folk memory occurred in Strathnaver in June 1814. Impatient with tenants who had ignored notices to flit to the coast to make room for sheep farms, Sellar led a group of men to throw them out. The houses, once the inhabitants and their moveable possessions had been put out of doors, were burned. A bedridden woman in her nineties was dragged from one house. Twenty-seven families were evicted in this manner in the first week. Donald Macleod, whose father was among those evicted, was later to write bitterly about these events: '. . . many deaths ensued from alarm, fatigue and cold . . . Some old men took to . . . wandering about in a state approaching to, or of absolute insanity . . . pregnant women were taken with premature labour, and several children died.'[8] Sellar was brought to trial for 'culpable homicide, real injury and oppression' but the Inverness jury found him not guilty. In November 1818 he resigned from his post as factor, and the Sutherland family, anxious now to wipe away the opprobrium their improvement policies was attracting, also sacked William Young. During the following years, Sellar kept a watchful eye on the affairs of the estate from his home on the farm of Syre that he rented in Strathnaver, and bombarded his successors as factors with advice on how to go about things.

Clearances continued in Sutherland but were more carefully handled. The new man in charge was James Loch. Born near Edinburgh in 1780, the son of a landowner who had to sell up to meet debts, Loch spent his formative years with relatives, the Adam family of improvement and architecture fame, and studied law at Edinburgh University. He joined the Speculative Society and took part in debates and discussions, another bright young talent in post-Enlightenment Lowland Scotland. He managed the Adam estates after his graduation and eventually, in 1812, accepted the post of commissioner of the Sutherland estates, which he compared to a little kingdom, at a salary of £1,000 a year. Loch entertained no revolutionary political ideas. He appears to have held no doubts about the social system in the Highlands and he was clear about what he had to do, taking a long-term view of development and holding that decisions, seen in the first instance as radical and unpopular, would prove the right ones in the long run. In his time he was attacked fiercely for what happened in Sutherland, but also praised and sought after for advice on estate management. He

The rugged Caithness coast at Badbea. The roots of the houses occupied by crofters evicted from the Langwell area to make way for Sir John Sinclair's Cheviot sheep can be distinguished on the sloping ground above the cliff top.

continually defended the policies of his employer although he seemed to be under no illusions about his own position. In 1835 he wrote to the widowed duchess of Sutherland: 'Someone said to me at Tongue how much the young Lord [the marquis of Stafford, destined to be the third duke] will say of his grandfather and grandmother's exertions and those who executed their intentions. I said don't flatter yourself that this will be so, he will be told how much better these things might have been done and so ever will it be the case, and we must all submit to this.'[9] Towards the end of his life Loch admitted that he had made mistakes: the sheep farms had been too large, he thought. During the 40 years he directed the affairs of the Sutherland estate, the acreage of arable land in the county increased from 13,420 to 32,237, new villages, roads and harbours were built, and the fishing industry advanced considerably. He was MP for the Wick burghs from 1830 to 1852. During the election campaign in which he finally lost the seat, he was followed through the streets of Wick by men carrying a model of a burned-out Highland cottage. Blackened roof-trees and crumbling stone walls remain the most vivid, popular image of the age of improvement in Sutherland.

The Clearances in the northern counties were a late phase in a social revolution that swept across much of Europe. For example, between 1758 and 1830, almost all of the common land in Denmark was enclosed and given to individual farmers. State decrees got rid of an open tillage system similar to runrig, but, unlike in Britain, the Danish land reforms did not oust the small landholders, the men with a few strips in common fields and grazing rights, but created instead a society of small farmers living on farms between 25 and 80 acres in extent.[10] As Denmark is well endowed with arable land, a better comparison with the hill country in the Highlands may be sought in Wales, where events followed a pattern similar to those in the Highlands: the enclosure of large areas of hill land deprived poor people of grazing rights and precipitated large-scale emigration.[11] Removal of people from the countryside in the Scottish Lowlands and on the southern side of the Moray Firth had taken place without as much ado, but in the north a number of factors coincided to ensure the Clearances would be remembered. The changes happened later than elsewhere when the behaviour of landowners was coming under more critical scrutiny than before, and newspapers were beginning to report events. In the Highlands, there was also a strong element of culture clash. The landlords and their servants were by and large Scots or English by background and language whereas the tenantry to whom they were issuing eviction notices were predominantly Gaelic-speaking and products of a different cultural milieu, one that had been developing in parallel to that of their Lowland neighbours but at a different pace and with different values. Repeatedly the evicted spoke of betrayal as they tried to understand and come to terms with a new social contract that placed value on individualism and economic production rather than the old commonwealth of the clan. The 1813 evictions in Kildonan were attended with unrest, and riots also occurred at Culrain in the spring of 1820. Some 300–400 tenants, 'chiefly women', attacked 25 men of the Ross-shire Militia and 40 constables, pelting them with sticks and stones and forcing them to retreat. The sheriff was injured and had the panels of his carriage broken. The soldiers opened fire, wounding a few of the women. More trouble erupted in the following year.

It is remarkable that there was not more civil unrest in the early 1800s: the number of people enduring poverty was increasing and some of the promise of the glorious dreams of improvement was being shattered by outside economic forces. The end of the Napoleonic wars brought new economic factors into play. The first Sheep and Wool Market, destined to grow into a barometer of Highland agriculture, was held in June 1817 in Inverness. By this time the flocks of Cheviots and Blackfaces had become well established; it was estimated that there were 100,000 head of the former in Sutherland alone. In 1818 wool prices jumped to £2 per stone and it was also a good and early harvest. After that, however, wool prices slumped and did not reach £2 a stone again for many decades. Sheep prices held up better but they too fell, sometimes as in 1824 to half the price fetched in 1818. The fall in prices was also seen in the cattle trade. The great changes in the economy that were pushed through during this period were financed from various sources. A few landowners such as the wealthy marquis of Stafford had their own deep pockets in which to dip and were able to absorb sizeable losses. Others borrowed, and hoped their plans would bear fruit. The British Fisheries Society, to whom we shall turn below, received grants from the government to supplement funds raised from individual subscribers, and fish-curers and skippers were encouraged by government-run bounty schemes. Considerable pump-priming funds also flowed into Scotland from the Empire, especially in the case of the north from sons who made good in the Far East or in the slave plantations of the Caribbean.[12]

The poor relied on charity from their neighbours and from official sources, almost always the kirk, to keep body and soul together. Money taken in by the kirk session from the Sabbath collections, from benefactions and donations, and from the small fines imposed on parishioners found guilty of sin, was divided among the unfortunates on the parish poor roll. The amounts to be distributed were frequently small and became increasingly inadequate in the early 1800s. In January 1813 the magistrates and town council of Inverness decided to grant a premium of one shilling for every boll of meal brought into the town market over the following six months to relieve 'the distresses of the lower classes'. The poor roll for the burgh is given as 222 in the OSA, about 4 per cent of the population but their numbers increased in the following years and in 1815 a society was established to suppress begging. A soup kitchen was opened two years later, and this was followed by a distribution of coal to some 700 people, all done by charitable efforts. The other burghs went through a similar experience. The situation was made worse by the eruption of Tambora in 1815 in what is now Indonesia, a volcanic event that affected the atmosphere around the northern hemisphere and ruined harvests across Europe in 1816. Elgin raised £300 in the hope that the poor could be provided for until the harvest could offer employment. The marquis of Huntly obtained 600 bolls of meal from the government to relieve dearth on his estates; the laird of Ballindalloch bought 800 bolls in Banffshire and 500 more from Berwick to supply his tenants; and the marquis of Stafford is reported to have spent over £7,000 on 3,400 bolls of meal and 500 bolls of seed potatoes for his tenants, as well as buying cattle. Other landowners were cutting rents or buying oats on credit from the government. Some measures were more exotic: we do not know what the tenants of Mr Grant of Elchies thought of their laird shipping in 3,000 pounds of rice and 200 hundredweights of treacle.

It says much for the character of the majority of the people that they endured their predicament stoically. In May 1817 a man in Nairn was sentenced to 14 years' transportation for sheep-stealing but fear of such draconian retribution was not needed to prevent most from lawbreaking. One rural activity that kept many cottagers from starvation and which at the same time attracted vigorous law enforcement was smuggling. In 1824, indeed, an Inverness newspaper suggested that landowners had a vested interest in the illicit distilling of whisky; they could let otherwise useless pieces of moor to smugglers for several hundred pounds and then turn a blind eye to the thin spirals of smoke rising from the howffs. Glenlivet was a major centre for smuggling. Large fines, however, began to frighten off some of the organised smugglers: 130 people were found guilty at a trial in Tain in 1825 and sentenced to pay fines of up to £100. In March 1828 the 20 people in Inverness prison for smuggling included four women, one of whom was over 70 years of age. Evicted tenants were under the strongest threat of destitution, their poverty compounded by the psychological shock of having been driven from the places they knew. Squatting where it was possible to delve a few drills of potatoes became common. In Kilmuir Easter evictees crowded on to waste ground where they bought an acre or two on which the proprietor allowed them to stay for seven years in exchange for a token rent in hens and eggs. In Kiltarlity the dispossessed made crofts for themselves on the moors. Colin Mackenzie of Kilcoy on the Black Isle made room on his portion of the former common land of Millbuie for tenants evicted from Redcastle and for 'strangers expelled from various parts of the Highlands'.[13] There was a considerable movement of people from Kildonan and Strathnaver into the Caithness parish of Latheron, doubling the population by 1840 when the Revd George Davidson wrote his contribution for the *New Statistical Account of Scotland*.[14] Close to 500 evicted from Assynt and Strathnaver settled in the parish of Dunnet in 1821 but, recorded Thomas Jolly severely, 'Their habits not being adapted to an industrious life, they soon got in arrears with the landlord, and went off, some to the Highlands, others to America'.[15] James Loch recorded 408 families squatting in Sutherland in 1815–16 and said that they were fugitives from clearances in Ross-shire. In July 1817 he found people starving and destitute sleeping on the beaches along the Firth so that they could beg from fishing boats and, unable to stand the sight, ordered a distribution of free food.

Emigration – to the south, or overseas – became the resort of many, and in the early decades of the nineteenth century the departure of emigrant ships from the ports of the Firth became a frequent occurrence. The voyage across the Atlantic took on average around 40 days but it could drag out to three months, an endurance test of overcrowding, rats, fetid water, disease and bad food for people who often had little knowledge of the sea. Ignorance of a less honest society than the one they came from also opened them to being duped. The Sutherland Transatlantic Friendly Society was formed by one Thomas Dudgeon of Fearn in 1819 at a public meeting at Meikle Ferry attended by over 1,000 people, many of whom there and then parted with sixpence or a shilling as a subscription. Dudgeon's oratory, full of promise for his followers and laden with wrath against the Staffords and their factors, moved the authorities to act; his subsequent meetings were declared illegal, he was labelled a crook, the society was disbanded and the hopeful subscribers lost their money. Better organised were the schemes of Thomas Douglas, Lord Selkirk, who in 1813 took a batch of emigrants from Kildonan to his new settlement on Red River, Manitoba, the first of over 1,000 people to complete that migration.

The predicament in which the rural poor, especially the evicted and displaced, found themselves might have been much worse had it not been for the massive expansion of the herring fishery in the early 1800s. Large Dutch herring fleets, made up of busses – in effect factory ships – and smaller boats for catching, had been coming to the Scottish coast every year for a long time before organised attempts were made at home to build up exploitation of this resource. The last and, in the event, the most momentous attempt began with an Act of the government in 1786 that brought into being the portentously titled British Society for Extending the Fisheries and Improving the Sea Coasts of This Kingdom – to become known in time as the British Fisheries Society (BFS). The same Act modified an already existing system of bounty payments, in effect subsidies, to encourage the participation in the fishery of inshore fishing boats. The BFS established three planned settlements on the west coast, at Tobermory, Lochbay and Ullapool, before turning its attention to the opposite shore, despatching the engineer Thomas Telford on a tour to find a suitable site for a new herring port. On the east coast of Caithness, Telford visited all the existing creeks and havens before deciding on Wick as the place where the BFS should focus its efforts. The town was then a large village, rows of buildings straggling along the north side of the Wick river. The large parish church dominated the west end of the main street and a wooden bridge spanned the river to join the road heading south. The land belonged to Sir Benjamin Dunbar of Hempriggs, who had already established a small fishing settlement at the east end of the town and named it Louisburgh after his wife. The BFS promised development on a scale only dreamed of hitherto, but it was a dream that took several years to become reality. It was not until 1801, over a decade since Telford's visit, that Sir John Sinclair, in one of his many roles in development, this time as a director of the BFS, opened discussion with Alexander Miller, merchant, entrepreneur and Sir Benjamin's man of affairs on the spot in Wick, over the acquisition of the south bank of the river for the new fishing settlement. At the time, Sir Benjamin was on service with his regiment of fencibles in Ireland, necessitating much writing of letters back and forth before signatures were at last appended to the final agreement on 11 March 1803.

The BFS initiated construction in 1806, first replacing the wooden bridge with a stone one and then starting the building of the new harbour to the design of Thomas Telford and under the supervision of the local architect George Burn. The adjacent settlement with its grid of streets named after the directors of the BFS was called Pulteneytown in honour of the governor of the BFS, Sir William Pulteney, who had died in 1805. In little more than ten years its population had passed 850, including fish-curers, merchants, tradesmen, shopkeepers, contractors, and artisans and their families, and the catching and processing of herring was creating a growing wave of prosperity that would not peak for a generation or two. The glittering prospect of money to be made attracted fishing crews from

Opposite: At the height of the herring season in 1865, Wick harbour was crowded with fishing boats, while around the quays women laboured at the gutting troughs, cleaning the fish and packing them in barrels of brine.

around the Firth for the six-week summer season. In the first 20 years of its existence the catch of herring in Wick swelled from 10,000 to almost 200,000 barrels per year. Such riches ensured a fast spread of the fishery to all the ports and landing places around the coast. Helmsdale produced over 5,300 barrels of salt herring in 1815 and increased that total to 46,571 barrels by 1839. The coast from Gardenstown to Portsoy, the area under the eye of the fishery officer based in Banff, was producing around 30,000 barrels a year by the late 1830s. Every summer saw a great increase in traffic in the Firth between the south side and the Caithness coast, as boats made their way to Wick. This seaborne migration was accompanied by another from the west as workers flocked from the Hebrides and the west Highlands to find employment as fishers, labourers or gutters, packing into the grey streets of Pulteneytown, swelling the population of the place to three times its normal size. George Burn's harbour could no longer cope with the press of boats and in 1824 the BFS began to extend its sheltering arms. The minister in Buckie in 1842 was one of several observers who noted the basic economics of the fishery; before the season began, a skipper would engage to land all his catch to a fish-curer for a guaranteed £8–10 plus 10s 6d for each cran (a volume measure roughly equal to 1,000 herring) plus four pints of whisky each week. Over the season the skipper could hope to catch 100 cran, a landing that produced an income of £18,375 when expanded to the 245 boats that called the Buckie coast home.[16]

Just as the herring fishery was burgeoning, another great project under Thomas Telford's supervision was completed – the digging of the Caledonian Canal from Inverness to Fort William. This added another route to the trading network emanating from the Firth, making it easy to sail cargoes of fish to Dublin, Belfast and Cork. Roads in the far north, however, were still in a primitive state. The squads of soldiers labouring to drive roads from the south through the mountains to the Firth coast in the previous century had not plied their spades beyond the Black Isle. The east coast of Sutherland was served by 'a broken, rugged pathway, running by the seashore from the Ord Head to Meikle Ferry' as Donald Sage found when he rode from Kildonan to school in Dornoch in 1801.[17] The road over the Ord into Caithness was notorious among travellers, and a journey from the northernmost county to Edinburgh could take eight days. As might have been expected, Sir John Sinclair showed an interest in infrastructure: using the statute labour laws that decreed that able-bodied men could devote

six days a year to road maintenance, he assembled as many workers as he could and, in one day, goes the story, laid the route of the Causewaymire (now the last stretch of the A9) from Latheron to Thurso. The absence of bridges – there was only one in Sutherland, at Brora – meant fords had to be negotiated or a boat used. Regular ferries crossed at Little Ferry at the mouth of Loch Fleet, Meikle Ferry in the Dornoch Firth, between Cromarty and Nigg, and between North and South Kessock by Inverness. Accidents were frequent on the ferries, possibly the worst at Meikle Ferry in August 1809 when a southbound boat, overcrowded with passengers heading for the Lammas Fair in Tain, capsized and 99 people were drowned. A less tragic sinking 'completely ducked' the people returning to Beauly from the sacrament [communion service] at Kirkhill in June of the same year.

Thomas Telford was commissioned in 1801 by the government to survey the roads in the north. His report on their deplorable state, unfit for the purposes of civil life and commerce, led to him being appointed in 1803 chief engineer for both the Commission for Highland Roads and Bridges and the Commission for the Caledonian Canal. Telford proposed that the costs of the new roads be shared between government and the

An image of the first bridge across the Dornoch Firth at Bonar, built to a Telford design in 1811–12, is preserved on this monument beside a successor. Telford's structure was swept away in a flood in 1892. The present bridge is the third on the site and was opened in 1973.

local landowners, who after all would benefit greatly, a suggestion that was reluctantly accepted and became law in July 1803. There followed a delay as landowners quarrelled over how the payments should be shared among themselves, until the Inverness-shire lairds petitioned Parliament to levy the road money according to rental assessment; the Inverness Assessment Act became law in 1804, and was followed by similar acts for Sutherland, Ross-shire and Caithness. Over the next 18 years Telford and his fellow engineers laid a new transport network across the country, nearly 1,000 miles of new road, over 1,000 bridges, creating more or less the network that has served until the present. Telford chose a semi-literate Forres mason, John Mitchell, to be his general superintendent, and John's son, Joseph, followed his father in the post. Gradually the main north highway was pushed up the northern shore of the Firth, establishing the spine of the network. The Beauly River at Lovat and the Conon were bridged in 1810, the Dornoch Firth at Bonar in 1812, and the construction of the Mound to carry the road across the marshy headwaters of Loch Fleet was completed in 1816 (under the supervision of Young and Sellar). Along the new road came coaches and an improved postal service. The coach 'Duchess of Gordon' began regular runs between Perth and Inverness in 1809 and, in the same year, a diligence pioneered the route north to Tain, whence mail for points to the north and west was carried by a runner. The runners were doomed to be supplanted by the coaches but for decades this hardy bunch of men, on foot or on horseback, had kept open the lines of communication; a weekly runner at a salary of £20 a year was recommended to link Bonar Bridge with Assynt in 1828. A daily coach service began between Aberdeen and Inverness in 1811, replacing the post rider who had made the journey three times a week. In 1819 the mail coach opened the route to Thurso; the timetable required leaving Inverness at six in the morning and reaching the northern destination at close to noon on the following day, with the return journey departing from Thurso at seven in the evening and arriving in Inverness around the next midnight: a 159-mile, 28-hour ordeal with 12 changes of horses.

Coach services improved in time but it is hardly a surprise that those who could preferred to make long journeys by sea. Coastal shipping benefited from the surveying of better charts in the late 1700s and after the appointment of the Commissioners of Northern Lights in 1786 came lighthouses – on Kinnaird Head in 1787 and the Pentland Skerries in 1794. Several ports and villages around the Firth could boast their locally owned

trading vessels, only one or two in some instances but in the larger centres sizeable fleets. Nairn had seven vessels in trade in 1840 but Banff had one brig, 18 schooners and 48 sloops registered at its custom house, which covered the stretch of coast from the Spey to Fraserburgh. 'The exporting of live cattle to London was first tried, as a speculation, in 1826,' noted the Banff minister in 1841, 'and since that time has formed a regular branch of trade.'[18] Other livestock, grain and potatoes were also exported from Banff.

Coastal sailing vessels could often be easily beached on the sand and gravel in parts of the inner Firth but piers and havens became crucial to provide shelter and accommodate larger ships. Several harbours had been constructed or improved in the eighteenth century but now a major phase of expansion took place, under the direction of the indefatigable Thomas Telford and other engineers.

The A9 sweeps around the head of Loch Fleet, here at low tide mostly acres of gleaming mud and sand. The road lies on a great embankment, the Mound, built by Telford's workers. Behind it the partially impeded flow of the Fleet waters and an alder wood.

The early 1800s are also marked by a phenomenon that was akin to a rural enlightenment. Alongside the developments in the economy and a rise in the standard of living of a significant section of the population emerged a new interest in the environment, science, history and ideas. The foremost example of this is the career of Hugh Miller. Born in Cromarty in 1802, Miller left school to work as an apprentice stonemason in the quarry owned by his uncle where he was soon struck by the appearance in the local rock of the marks of ancient waves on a sandy shore. His curiosity and intellect stimulated, Miller went on to collect fossils and disprove the current opinion among geologists that Old Red Sandstone, the rock strata underlying much of the Firth coast, was barren of such remains. Miller developed his research, communicating with the most important scientists of his day, while retaining a deep religiosity that led him to struggle to reconcile geology with the accounts of creation presented in Genesis. He also became strongly involved in church politics during the Disruption, of which more below, and was appointed editor of the Free Church's newspaper *The Witness* before his death in 1854. Other men who had little formal education beyond basic schooling but followed an inner drive to pursue an intellectual interest were Robert Dick (1810–66), the Thurso baker who made significant discoveries in botany and geology, and Alexander Bain (1810–77) who rose from being a watchmaker's apprentice in Wick to invent the electric clock and other devices. Banff shoemaker Thomas Edward (1814–86) studied wild animals and practised taxidermy, and became a noted exhibitor of wildlife. A slightly earlier exemplar of this class of gifted men was James Ferguson (1710–76) from Banff who combined a mechanical genius with an interest in astronomy.

Newspapers began to appear regularly in the early 1800s, and from the beginning carried features on local history and antiquarian topics. Lending libraries and reading clubs also sprang into being and provided fodder for the enquiring mind. The parish library in Banff boasted 164 volumes on religion and general knowledge, available to members for a subscription of sixpence per half-year. Kincardine parish on the southern shore of the Dornoch Firth had a reading club with 16 subscribers in the 1830s where the secretary bought works of travel, biography and general literature every year. At around the same time, in Rosemarkie the minister noted a taste for reading in all classes but particularly for books 'of a religious character'.[19] A considerable proportion of the intellectual activity of the age went into religion. Apart from teaching, still a somewhat lowly regarded profession and certainly not too well paid, the obvious ambition for a 'lad of pairts' in the Firth region as in the rest of the country was to enter the ministry. This, combined with the custom of sons following fathers in their occupations, resulted in dynasties of clerics. Donald Sage, who was minister of Resolis in 1840 when he wrote *Memorabilia Domestica*, was the son and grandson of ministers, and the father of another, as well as being connected by marriage to several more. One folk memory of the Clearances distinguishes the ministers for having been craven tools in the hands of lairds and factors. This is not entirely true, and neither is it true that all the men of the cloth were zealous

Thomas Edward, the Banff shoemaker who became famous for his skills at taxidermy and his exhibitions of specimens of wildlife.

Calvinists who frowned on every sort of pleasure. When Donald Sage's father, Alexander, married for the second time in December 1794, the feasting and dancing at the Kildonan manse were open to all the sub-tenants and elders. Music was provided by the bagpipes, and great hilarity resulted when the minister's housekeeper dumped handfuls of flour on the bald head of an elder. On other occasions the fiddle accompanied reels and strathspeys at the manse, and a local poacher and smuggler was often called upon to make the household's annual 'brewst' of malt whisky. Many ministers were interested in secular topics. The incumbent in Rogart, Eneas Macleod, was a poetry enthusiast: a close friend of the Sutherland poet Rob Donn Mackay, he wrote down the compositions of this illiterate bard and saved them for posterity. We owe the preservation of much folklore to the ministers who recorded stories from their parishioners and the various contributions to the *New Statistical Account of Scotland*, of which more below, often show their clerical authors had a keen knowledge of local flora, fauna and geology.

The tenets of Christianity set the tone for daily life, and the Bible influenced speech and writing. The Revd George Mackay in Brora noted that his flock could 'quote scripture in support of their arguments with surprising readiness and accuracy' although they otherwise had little education.[20] This knowledge did not mean ministers always had an easy time, as it was not at all uncommon for laymen and elders, 'eminent for piety' in Donald Sage's phrase, to challenge their minister on sometimes obscure doctrinal issues. In rural areas conventional Christian beliefs sat alongside, apparently quite happily, superstitions lingering as echoes of distant medieval Christianity or even older Celtic beliefs. Although the vast bulk of the population belonged to the Kirk of Scotland, there were sizeable Roman Catholic minorities in a few corners of the Highlands. The Scottish Catholic Relief Act was not passed until 1793 but throughout the century a trickle of priests and seminarians had kept open the links to the Continental Scots colleges; a seminary existed at Scalan in a lonely corner of Glenlivet from 1715 until 1799, in which time it trained over 100 priests. Up in Caithness, Sir William Sinclair, the laird of Keiss, had further spiced the ecclesiastical brew by founding Scotland's first Baptist congregation in the 1750s. The Revd George Innes in Deskford in 1836 noted less drinking among his parishioners and more reverence for the Sabbath 'and the introduction of Sabbath schools and Sabbath school books has led to passing the other evenings of the week in a more improving manner'.[21] In the same period, in Forglen parish kirk the Bible class held before the regular morning service attracted 30 to 40 youngsters every Sunday, and a further 70 to 80 attended the Sunday school: 'It has studded the church with young faces,' wrote the minister approvingly.[22] In Mortlach the number of communicants, i.e. those taking communion, a sacrament open only to parishioners who met a certain standard in their understanding of the faith, was 'never less than 700', roughly a quarter of the population.[23]

Religious views spilled over into the secular sphere, influencing the people's reactions to social and political issues, and in return the Kirk paid close attention to the lives of its parishioners. Between 1834 and 1845 the General Assembly oversaw the production of a *New Statistical Account* of the country, updating and enlarging upon the volumes edited by Sir John Sinclair some 40 years before. The authors of the contributions to the NSA, as it is known, were in the main the parish ministers, and now they had occasion to record the momentous changes in the preceding decades, noting the great improvements but also lamenting the effects of dispossession on their people. The Revd David Carment in Rosskeen thought the depopulation caused by the formation of large farms 'a serious evil', wiping out an independent peasantry, degrading their morals and embittering their spirits.[24] The minister of Nigg stated that the improvements in agriculture had been carried through at too high a cost, that beautifying the face of the earth was wasted without preserving 'moral beauty' – 'the luxury of doing good, and the pleasure of being surrounded by a moral, a grateful and a happy population'.[25] Occasionally there appears in the NSA a defence of the Highlander, such as by Charles Downie in Contin who thought them 'grossly calumniated when represented as inactive and indolent' and felt they were the equal of any when given proper encouragement.[26] In Brora, George Mackay, the minister of the parish of Clyne, wrote of the virtues of the Highlanders and combined it with an apologia for the Clearances, depicting the people in the uplands as living in former times a simple, moral life but one vulnerable to dearth and illness when their only recourse was to appeal for the landlord's help, a state of affairs that 'could not continue while the rest of the world was . . . making rapid advances in improvement'.[27]

Despite the regrettable decline in moral behaviour – often synonymous with too much drinking, rising living standards were universally acknowledged to be happening around the Firth. In Loth the small tenants now had 'comfortable stone cottages of improved construction',[28] the villages were growing, waste ground was being brought into cultivation – by the time of the NSA there was little common land to be found anywhere although

Surf breaks on the beach at the mouth of the Deveron between Banff and Macduff. The Doune church on the skyline was built in 1805 and remodelled in 1865 in a distinctive Italianate style with a cupola.

The cupola on the farm steading complex of Conan Mains near Conon Bridge is a supreme manifestation of the competitive pride and enthusiasm of nineteenth-century lairds for agricultural improvement. The octagonal doocot can also be seen. The contrast with the plain functional steadings of today is starkly plain.

an undivided 600 acres still existed between Foulis and Inchcoulter, a canal had been cut to allow ships to come into Dingwall, notice was being taken of the archaeological relics near Portmahomack, and the towns were now boasting banks, gaslight and clubs. The public highway at Nairn, wrote the Revd James Grant, was 'Macadamd', no longer rough but inconveniently dusty in dry weather.

The increasing opportunities for education and rising standards of living encouraged interest in political reform. The final passing of the Great Reform Act in May 1832 was greeted with large public demonstrations. Some 4,000 people gathered in Banff from many villages along the coast, marching cheerfully under flags and banners, one of which read: 'From the rotten burgh system, Good Lord, deliver us'.[29] A bigger crowd flocked to the Academy Park in Inverness to attend what the *Inverness Courier* called 'the first open meeting ever held in Inverness for a political purpose'.[30] The inclusion of more of the population in a slowly widening electorate would continue for many decades into the future but, in contrast with this essentially secular process, the 1840s saw a major event that combined politics with religion – the Disruption of 1843.

For a long time the Church of Scotland, the Kirk, had harboured a split over the issue of patronage, over who had the right to choose the minister, the local patron of the kirk – usually the laird – or the congregation. The right of the congregation stemmed from the Reformation but its implementation had become eroded over the years until in 1834 it was reinforced when the General Assembly of the Kirk passed the Veto Act. In the words of the Revd Thomas Brown, who wrote the momentous *Annals of the Disruption*, 'Unacceptable ministers were no longer to be thrust on unwilling congregations'.[31] The first test case, at Auchterarder where the minister chosen by the laird, Lord Kinnoul, was rejected by the people, resulted in the House of Lords backing the Court of Session and overruling the constitution of the Kirk. The next test case occurred in Marnoch in Strathbogie where, amid January snow in 1841, the congregation rose in a body and walked out on their new minister. It was as if the people, denied much say in political affairs, decided to revolt in a matter of religious conscience. The climax of the dispute came at the General Assembly on 18 May 1843 when 400 ministers left the Church of Scotland to form the new Free Church. To leave the established church was no easy decision to make for ministers with families, as the 'rebels' had to quit their manses and livings until the new church was able to acquire and erect new buildings of its own. The long-term result of the Disruption was the influential presence of the Free Church throughout the north. Despite the strength of feeling on both sides, the massive schism was free from violence although one incident veered riskily close to it – at Resolis in 1843, young men gathered to prevent the 'intrusion' of a new minister after Donald Sage had left to join the Free Church, but found their way barred by a company of soldiers from Fort George who fired over their heads, first with blanks and then with ball until some order was restored. Later that day, a mob stormed into Cromarty to free a woman, arrested for cheering on the crowd at Resolis, from the jail.

The great flood of Moray in 1829 was a reminder that nature was not always biddable. In that year, a hot, dry spell came to an abrupt end in early August with extremely heavy rain in Caithness and the Monadliath hills. The Nairn, Findhorn, Lossie, Spey and Deveron, with all their tributaries, were roused to fury and burst their banks, destroying buildings, bridges and crops. In all the devastation only eight human lives were lost, a happenstance Sir Thomas Dick Lauder in his account of the catastrophe

The small village of Rockfield is the most northerly of the four seaboard villages of Easter Ross. The others are Shandwick, Balintore and Hilton. In 1855, sixteen fishing boats worked from Rockfield. This picture also shows well the ancient raised beach that runs along the Firth coast to remind us of the much higher sea level some 8,000 years ago.

ascribed to 'Providential deliverance'.[32] Barely three weeks after that, more downpours struck the mountains west of the Great Glen, sending a flood down the rivers that had escaped spate the first time. In January 1849 another great flood occurred in the Great Glen, sweeping away the stone bridge at Inverness and inundating all the low-lying parts of the town. The rivers draining into the Moray Firth remain prone to spate and flooding.

In 1831 reports began to appear in the local papers of an epidemic of cholera in central Europe and as the year passed this threat crept closer until in November local authorities up and down the land began to take steps to counter it. Inverness Town Council, for example, set up a board of health with subordinate district committees to handle the emergency. Forres town council borrowed £100 to build a cholera hospital. In May 1832, two cases were reported among the soldiers at Fort George but the full outbreak did not begin until August, significantly at the time of the herring season when all was bustle. Helmsdale was the point through which the disease arrived. The *Inverness Courier* reported what happened: '. . . a boat from Prestonpans, in the Firth of Forth, arrived . . . the all-important fact that one of the crew had died of cholera during the voyage, being either concealed or overlooked, the survivors were most unfortunately permitted to land and to associate without any restraint with the healthy population of the place'.[33] The disease spread north and south, leapfrogging through the crowded herring ports, within four weeks infecting the coast from Wick round to Nairn. In Cromarty, the people took the law into their own hands and imposed a quarantine, admitting no strangers. By the end of October there had been 553 cases in Inverness, with 175 deaths. The extent of the outbreak varied greatly from one community to the next: Wick had 66 deaths, Golspie only three, yet one-third of the inhabitants of the village of Inver perished. The death toll from cholera was highest in the crowded, unsanitary slums in the southern cities, especially in Glasgow. The Firth region escaped relatively lightly, protected to some degree by its isolation and the largely rural distribution of the population. Perhaps the outbreaks of plague in the Middle Ages about which we know so little followed a similar course to the one taken by the 'cholera morbus', in this sense making the event of 1832 arguably the last of the old-style epidemics before improved sanitation and water supplies altered public health for the better.

By the 1840s the potato had become the staple diet of a large section of the population. The minister of Dornoch said it was the chief subsistence of the poor for a half to two-thirds of the year. The appearance in the country of the potato blight in 1846 therefore threatened famine, a visitation of a type of disaster that many may have regarded as a threat safely confined to past ages. The crop wizened in the field or, worse, even if harvested turned black and rotten in the storage pits, the effect of the fungus *Phytophthora infestans*, not understood at the time. The potato blight in the Highlands had nothing like the catastrophic consequences in Ireland where large numbers of the rural poor died from want, but it did weaken people, reduce them to desperate poverty and make them susceptible to illnesses associated with malnutrition. It also provoked food riots in several localities around the Firth, usually in angry reaction to the sight of ships loading local crops for export. In Inverness in February 1846, a large, angry crowd defied the magistrates, special constables and units from the 87th Regiment to prevent two ships loading potatoes. In February 1847, a crowd of people from Evanton gathered on the beach at Foulis and turned back wagons on their way to load waiting ships. There were other incidents in various places, including Avoch, Burghead, Forres and Macduff, much to the outrage of

The makeover of Dunrobin Castle by Charles Barry in the 1850s for the second duke of Sutherland masks the thirteenth-century keep in its core.

the farmers whose trade was interrupted, and several times troops were called out in an effort to maintain public order. Two companies of the 76th Regiment fired on rioters in Wick, killing a young girl. Bayonet charges were required to disperse rioters in Invergordon. The last food riot in the region was probably the one in Banff in 1874, when locals blocked the harbour to prevent a ship sailing with a cargo of meal.

The coming of the railway in mid century altered the commercial geography of the region and had far-reaching consequences. For most of the Moray Firth area, it sealed the route through Strathspey and over the Drumochter Pass as the main connection with the south, eclipsing the old way skirting the east of the Cairngorms. Several companies vied to create the new transport links and in the struggle some were subsumed within their larger rivals. The Great North of Scotland Railway (the GNSR) and the Highland Railway, with its headquarters in Inverness, were the two main competitors. In September 1854 the GNSR opened its line from Aberdeen to Huntly, barely three days before work began on the independent Inverness–Nairn connection. The rails from Inverness reached Nairn in 1855, Dingwall in 1862, Bonar Bridge two years later, and finally Wick and Thurso in 1874.

The GNSR line extended from Huntly to Keith by October 1856, two years before the Inverness–Aberdeen Junction Railway pushed east to the same destination. GNSR sought to lay its own route through to Inverness but Elgin was as far as it got. A line from Forres via Strathspey to Dunkeld was finished in 1863 by the Inverness and Perth Junction Railway, but it took until November 1898 to complete the final stretches of the direct Perth–Inverness route. In that time most of the towns and villages around the Firth became bound to each other by the miles of iron track. The engineers and navvies had to overcome the landscape to build the rail lines, not so difficult a task through the lowlands of Moray but a daunting undertaking elsewhere, with the viaducts over the Findhorn and Nairn rivers and the valley at Cullen forming notable tributes to their labour. Branch lines were built to connect to villages off the main routes – to all the centres between Burghead and Fraserburgh on the southern shore of the Firth, and to Fortrose, Dornoch and Lybster on the northern side. To be bypassed by the rails meant economic decline – Findhorn lost its role as Moray's principal port and, despite efforts to establish a rail connection, Cromarty remained with only dwindling maritime links until a new road network in the 1980s put an end to its isolation. The duke of Sutherland, seeking to encourage

The official opening ceremony of the duke of Sutherland's railway on 1 November 1870.

The schooner *Elba* loading with barrels of salt herring at Wick, c.1890.

economic development in the north-west, had the line make a sweep inland to Lairg, a detour that did little in the long term to counter the difficulties imposed by climate and rock.

The herring season continued to be a highlight of the calendar and by the 1860s, when the fishery peaked in Wick, was attracting tourists to marvel at the bustle of some 5,000 fishermen, backed by almost as many gutters, packers, coopers and carters on shore, landing and processing the seemingly endless flow of the 'silver darlings'. The growth of the fishery was marked by the maintenance of quality standards for the barrelled herring, the introduction of cotton nets to replace the heavy hemp of the early years, improvements in boat design and harbours, and the growth of trade networks. Despite the occasional bad season the growth continued right through the nineteenth century and did not peak until the eve of the First

World War when annual production was around 2,000,000 barrels.[34] By then, most of this product was exported to Continental Europe. Walter Biggar, an Edinburgh merchant who married a Banff woman and settled there in 1821, is credited with starting the trade to the Baltic. In fact it was already under way from other ports but, judging from the shipping lists in the newspapers, the Baltic remained a relatively minor export destination until around 1840. Ireland and, via English ports, the slave plantations of the West Indies were more important markets in the early decades. By 1849, though, the harbour master at Wick was able to record in his log, for example on 7 September, 'A great quantity of herrings has been shipped to the Continental markets.'[35] During the rest of the 1800s and into the twentieth century Danzig (now Gdansk), Stettin (now Szczecin), Hamburg, Königsberg (now Kaliningrad), Memel (now Klaipeda) and many other northern harbours became regular destinations for the Firth seamen. The local papers regularly carried exchange rates for the rouble and the mark, and reports from eastern Europe on anything that could affect the trade, while the fish-curers in Wick, Buckie, Banff and Macduff

The railway viaduct at Cullen, built in 1882–84, spans the deep valley of the burn at the west end of the Seatown, its jumble of roofs easily distinguished from the more formal grid favoured by town planners. Strictly speaking there was no need for the viaduct to be built but the lairds of Seafield refused to allow the railway to pass along an easier route too close to their residence.

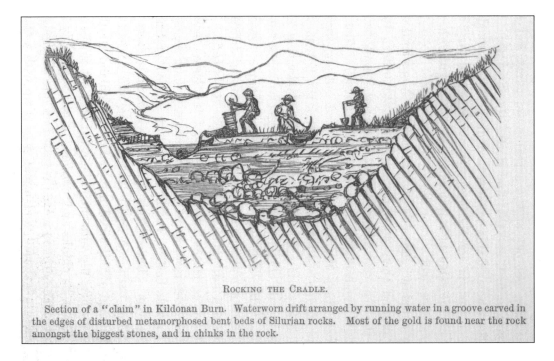

ROCKING THE CRADLE.

Section of a "claim" in Kildonan Burn. Waterworn drift arranged by running water in a groove carved in the edges of disturbed metamorphosed bent beds of Silurian rocks. Most of the gold is found near the rock amongst the biggest stones, and in chinks in the rock.

Hopeful prospectors hard at work in the search for gold on the Kildonan diggings in 1868–1869.

planned the building of villas to display how kind the sea and providence had been to them.

In pursuit of the herring, the fishermen built larger boats and experiments with design led to the appearance of the zulu, so named because it first sailed at the time of the Zulu Wars in 1879, a hull shape that combined the best features of its ancestors, the scaffie and the fifie, and gave crucial advantage in the race to bring a fresh catch to the quay to claim the best price. The sea and the fishing supported ship and boatbuilding, rope and net manufacture and ancillary trades in all the coastal towns, while inland agriculture could continue only with the underpinning of a thriving range of small village industries concerned with the making and mending of agricultural implements. All of these activities, with quarrying and some small-scale foundry work, comprised the extent of industry around the Firth. The economy remained rooted in the winning and processing of the raw produce of the sea and the soil. Forestry was confined to a few private estates. Many of the textile ventures established during the previous century failed, outdone by the cheaper products of the factories now spreading through the southern parts of Britain. Portsoy enjoyed moderate renown as a source of serpentine, also known as Scottish marble, but the search for useful minerals elsewhere in the region found only small deposits or pockets of limestone that could be burned for local fertiliser. There was a short-lived rush for gold in Strath Halladale in 1868–69. Brora was exceptional with its coal mine and its brickworks, as was Inverness where the Highland Railway company built locomotives.

The railway brought an end to cattle droving, and the mart where livestock was sold by auction and shipped out became a standard feature of country life. Another consequence of the railway was the much improved access of the north to southern visitors. Fired with enthusiasm for the wild scenery of the Highlands by James Macpherson's Ossian poems, Sir Walter Scott's romantic novels and tartan-bedecked arrangements for the state visit of George IV, and the love for Deeside displayed by Victoria and Albert, the north became a playground for the aristocracy and for a southern moneyed class with increased leisure time and a desire to be fashionable. The immediate coastal country around the Firth was not, however, what this new tide of tourists wanted. The cornfields of Moray and Easter Ross and the small towns and villages were and remain attractive but, although the Spey and its neighbouring rivers along with the heather slopes of the Cairngorm foothills provided plenty of sport for the hunting set, the focus of tourism lay to the west along the glens and bens of the 'true' Highlands. Inverness benefited from this new industry but not so much Elgin, Forres and other points east. The guidebooks reflect this: *Blackie's Picturesque Tourist of Scotland* for 1867 devotes over 70 pages to Perthshire, some 60 to Argyll and the west coast, and 35 to Skye and Inverness but races through Banffshire and Morayshire – or Elginshire, as it then was – in only ten.

The bed of the old railway line from Muir of Ord to Fortrose makes a good footpath.

Farmworkers in Moray.

The late 1800s saw the setting of many patterns that were to last into the present and shape life around the Firth for several generations. The large fermtouns had their often elaborate steadings and their hierarchies of grieve, ploughmen, cattlemen, dairymaids and other workers, while crofters struggled on their own to win a living from more thrawn upland soil. The Napier Commission gathered evidence of landlord oppression throughout the Highlands and its findings informed the 1886 Crofters' Holdings (Scotland) Act that brought about several reforms including security of tenure, although Moray, Banff and Buchan, officially deemed not to be 'crofting counties', were excluded from the legislation.

Towards the end of the Victorian period, along the coast the distinctive fishing communities followed their own rhythms of tide and season, and the towns and large villages flourished on a sense of their own significance. The professional classes of solicitors, teachers and businessmen now found new entertainments, indulging in organised sports – tennis, bowls, golf – and exploring their local environment on the new-fangled bicycle. The Inverness Scientific Society and Field Club was born in 1875, the Banffshire Field Club in 1880. The team sports of football and to a much lesser extent

The peculiarities of local government in mid-Victorian times: the great county of Ross stretches from the Moray Firth to the Atlantic, with Cromartyshire buried within it in a series of enclaves. The Black Isle also hosts an enclave of Nairnshire.

cricket and rugby also became organised into leagues. The Reform Acts of 1868 and 1884 extended the franchise to about two-thirds of the adult male population. The Education Act of 1872 brought in compulsory elementary schooling for all youngsters, administered by elected school boards of local worthies. In 1889 a local government act created county councils to operate along with the existing burgh councils, setting the boundaries of the counties that would endure for almost another century. This brought an end to the anomalous scatter of enclaves of land: the bulk of the old Cromartyshire comprised patches of territory across the body of Ross, and these now became one jurisdiction, the county of Ross and Cromarty.

# CHAPTER 12
# ROOTS OF THE FUTURE

The monument in Beauly to the raising of the Lovat Scouts for service in the Boer War.

In January 1900 the *People's Journal* carried a letter from Lord Lovat. 'I have been authorised by the War Office,' he wrote, 'to raise a corps of 180 stalkers, ghillies and picked men for, primarily, scouting services in South Africa . . .'[1] Thus was born the Lovat Scouts, the latest in a long line of regiments from the north of Scotland. As chief of the Frasers and the laird of one of the largest estates in the region, Lovat was falling back on the authority conferred by tradition. Barely 150 years before him, a direct ancestor had raised the 78th Highlanders to fight for the British cause in the Seven Years' War – they shared in the capture of Quebec from the French – and thereby redeem the family fortunes after the unfortunate choice of the wrong side in the last Jacobite rising. The export of fighting men from Scotland can be traced back to the Middle Ages and the Auld Alliance with France, but the post-Jacobite formations recruited to serve the Union Jack represented a major cultural shift with the adoption of tartans, kilts and bagpipes into the mainstream military, artefacts once associated with the 'barbaric' Highlander, the recent enemy in the Jacobite period. The battles with the French, from Canada to Waterloo, made the kilted Scottish soldier an icon, an image reinforced by stirring deeds in later conflicts around the Empire. This fed back to render the home society conscious of soldiering and to make things military firmly part of the Scottish identity, as can be seen in the contemporary newspapers. In 1900 Lovat's letter appeared amid reports and illustrations from the war against the Boers, where Sir Hector Macdonald (1857–1903) was in command of the Highland Brigade. Born the son of a crofter on the Black Isle, Sir Hector had quit an apprenticeship as a draper in Inverness to become a soldier in various theatres of the Empire, rising in the process from private to high rank, a local hero and an example to every ambitious lad. Pride in the role the country was playing throughout the world, in the form of its soldiers, emigrants and administrators, was strong. Regular reports on the dispositions of the Scottish regiments emphasised the intimate linking of national identity with Empire: for example, on 19 May 1900, the readers of the *People's Journal* learned that battalions of the Queen's Own Cameron Highlanders, the Gordons and the Seaforths were variously in South Africa, Gibraltar and Egypt, or in their home bases at Inverness, Aberdeen or Dingwall. For the Moray Firth people, these were the three home regiments, the three who more or less divided the region between them, the three in which local aspirants were most likely to enlist.

Throughout the nineteenth century, despite the swelling interest in things Celtic, Gaelic as a living language continued to decline. The contributors to the *New Statistical Account* in the 1840s wrote about this, noting the rapid progress of English in formerly strong Gaelic-speaking areas such as Easter Ross and the Sutherland coast. The Revd George Davidson, the minister of Latheron, went so far as to note a linguistic boundary, the Burn of

The career of Sir Hector Macdonald is outlined in stone on this tower erected to his memory at Mulbuie on the Black Isle, close to his birthplace.

This plaque inside Fort George is a reminder of the huge losses suffered by only one regiment during the First World War.

East Clyth, in Caithness between Gaelic and the local dialect of Scots. At the time of the 1891 census, Gaelic was still spoken in the upland parishes of Nairn, Moray and Banff and was similarly present in the western Caithness parishes. In contrast to the decline of Gaelic, the varieties of Scots spoken around the Firth showed signs of cultural vigour, with a dialect literature springing up in the columns of local papers and in books, developments that helped to shape strong local identities.

The dominating events of the twentieth century in the collective memory are probably the two World Wars. The outbreak of the First World War in August 1914 was marked by a surge in volunteering to fight in the country's cause against Germany. In September around 100 men were coming forward every day to join the Cameron Highlanders in Inverness and the other regiments were also attracting recruits. At the same time, as a harbinger of the grim years ahead, casualty lists began to appear in the local papers. The fighting in France settled into the bloody stalemate of trench warfare and it became clear that the war would certainly not be over before Christmas, as pundits were declaring back in the late summer. Gradually more and more of the nation's life became geared towards the war effort. Civilians organised themselves to support the troops. Rationing of commodities was introduced. In June 1915, under the legislation of the

Defence of the Realm Acts, nicknamed DORA although not with affection, Inverness and the northern Highlands became a restricted area, as part of the defensive cordon thrown around the naval bases at Scapa Flow and Invergordon. Compulsory conscription was introduced in 1916 for men aged 19 to 30, and all the while the casualty lists, the rolls of honour, filled the columns of the papers right up until the Armistice arrived at last in November 1918. Overall, some half a million men, about one quarter of the adult male population of Scotland, joined the armed forces, and around one in four did not return. The loss of such a substantial number of young men from all classes was the largest single effect of the conflict but the War also brought women into the workforce in larger numbers than before, a more prominent social change in the urban Lowlands than in the rural north where women had always laboured on the land.

The early decades of the century were also when many innovations appeared. The telephone had become widely familiar before 1900, at least in towns, but now there were new and dramatic forms of transport in the forms of the motor car and aircraft, and new entertainment in the cinema. The wireless appeared in the early 1920s. The folk in Inverness were given an early and local introduction to jazz when the US Navy set up a base in the town to service the great minefield laid across the North Sea in the last months of the War to contain the U-boat threat. The temperance movement which had been growing steadily in the Victorian years to combat the evils of drink had a boost and in the early 1920s referenda took place in some local government wards, depending on the number of signatures in favour, on whether or not to prohibit sales of alcohol altogether. Inverness rejected prohibition whereas Wick and Dingwall went 'dry', a decision that produced some interesting social consequences, not least a rise in illegal distilling.

The experiences of the War probably hastened the decline in interest in religion and the erosion of Sabbath observance, certainly in the towns and probably in the countryside as well, although a notable exception to this trend was an evangelical revival among fishermen in 1922. The rejection of organised religion was partly a reaction against the pre-War authority of the kirk and was tied up with a deepening awareness of politics and class among working people. In Scotland this is most often illustrated by the story of 'Red Clydeside' when strikes in Glasgow led to a fear among the authorities in 1919 of a Bolshevik-style uprising, a story that usually neglects to mention that hardly anything similar took place outside that city. There were strikes in Inverness among some workers but these were peaceable affairs and by and large the population around the Moray Firth regarded Glaswegian radicalism with suspicion. Even when unemployment rose to unprecedented levels in the 1920s and 1930s, and soup kitchens appeared to feed hungry bairns, dissent did not extend beyond grumbling. In fact, election results showed a rightward rather than a leftward trend. Electoral reform at the end of the War extended the right to vote to all adult men and to all women over the age of 30. Constituencies were revised: in came Banffshire, Caithness and Sutherland, Inverness-shire, Moray and Nairnshire, and Ross and Cromarty, with Buchan included in Aberdeen and Kincardine; and out went the old rural constituencies and the odd arrangement of groups of burghs. (From 1832 until 1918, Elgin, Banff, Cullen, Inverurie, Kintore and Peterhead had formed a single constituency, as had Inverness, Forres, Fortrose and Nairn; and Wick, Cromarty, Dingwall, Dornoch, Kirkwall and Tain.) In the so-called 'Khaki' election in December 1918, all the Firth seats were won by candidates of the Coalition Liberal party. The region stayed largely Liberal in its politics for the next 20 years but a swing to the political right became apparent in the 1930s, when the three constituencies from Nairn round to Aberdeen all returned Conservative Party members. The Labour Party made a respectable showing at times but no breakthrough. Ramsay MacDonald, Britain's first Labour prime minister, may have been born in Lossiemouth in 1866 but this brought him little political advantage in his birthplace.

As it happened, in 1926, the year of the General Strike, the Scottish Trades Union Congress held its 29th annual congress in the Wesleyan Central Hall in Inverness. At the opening reception laid on by the town council, Joseph Duncan, president of the STUC General Council, made his audience laugh when he described how the councillors had been less than welcoming 20 years before when he had come north as a 'propagandist'.[2] A month later, under the impact of the General Strike, supplies of merchandise and rail and steamer services were either delayed or suspended, and workers in several industries – transport, printing, iron and steel foundries, gas and electricity services, and construction – withdrew their labour. The strike lasted two weeks. *Inverness Courier* noted that the strikers, when they had gathered on the streets, had been almost entirely law-abiding, and claimed that 90 per cent of the men had come out against their will.[3]

The early 1930s, a time of high unemployment and considerable social distress, also saw the re-emergence of the nationalist movement; Home Rule had almost made it through Westminster in 1914 but had been laid

aside during the War. The National Party of Scotland put up five candidates in the General Election in October 1931, with John MacCormick, one of its leading figures, standing in Inverness. In the event, the voters of the Highland capital stuck with their popular Liberal incumbent, Sir Murdoch Macdonald, confining MacCormick to third place narrowly behind Labour. Inverness was home to other prominent nationalists, including the provost Alexander MacEwan and the novelist Neil Gunn, and it was in the latter's kitchen that representatives of the National Party of Scotland and its more right-wing rival, the Scottish Party, agreed in 1934 to lay aside their differences and unite.

The mouth of Dunbeath Water and the harbour of Dunbeath. The village was fictionalised as Dunster by Neil Gunn in his epic novel of the herring fishing, *The Silver Darlings* (1941). Gunn was born here in 1891. Dunbeath Water itself shaped his 1937 novel of childhood and youth, *Highland River*.

At the end of 1919 the Forestry Commission was established to develop a new timber industry. A few landowners such as Simon Fraser, Lord Lovat, had argued strongly for the development of state forestry before 1914 but it took the four years of warfare to show just how dangerously much the country depended on imports of timber. New forests of fast growing conifers were planted throughout the 1920s and 1930s to supplement private woodlands. One noticeable achievement of the Commission foresters was to tame the shifting dunes of the Culbin sands, to this day a thriving forest. High hopes were pinned to forestry to provide employment in rural areas and help retain people in the glens. Life on the old fermtouns was also changing. The machines were coming in, cutting the need for muscle power. As prices fell in the economic downturns, farmers, whether tenants or owners, had to struggle and for many on marginal ground it was too much, too tempting to abandon the land and find an easier, more secure living in the towns and villages. Aspirations were changing as well. The wireless and magazines taught children on the more remote crofts that better opportunities beckoned elsewhere. The districts around the Firth, especially in the more upland parts, are dotted with crumbling steadings, the reminders of evictions not through direct clearance but through the steady pressure of economic forces. The herring fishing continued to feature in the maritime calendar but also ran into hard times in the 1920s. The drifters enjoyed a short-lived boom immediately after the War, catching fish from stocks that for four years had had a rest from exploitation, but prices slumped in 1921 and fishing suffered along with the other sectors of the economy. The market in Europe for salt herring almost disappeared with the rampant inflation in Germany coming on the back of the Bolshevik revolution in Russia, and the coal strike in 1926 deprived the steam drifters of fuel. In Lossiemouth, Buckie and Macduff the fishermen found hope in a new method, the seine net, targeting bottom-dwelling and free-swimming white fish. This type of fishing gradually grew more popular during the 1920s and 1930s and, as the herring fishery slowly declined, became the dominant pursuit of the larger Firth boats, with the smaller inshore craft holding to the time-tried methods of line and creel.

The 1920s and 1930s were also a period of marked emigration. The population figures recorded for the censuses in 1911, 1921 and 1931 show a fall in numbers in all the counties bordering the Firth, a fall partly due of course to the losses during the First World War but also stemming from movement of people to the Scottish cities, England or the overseas dominions. The decline was not equally distributed. Caithness, Ross and Cromarty, and Sutherland lost proportionally more than Banff, Inverness, Moray and Nairn. Within the counties, the population figures for different parishes show that rural and upland parishes lost a larger percentage of their people between 1911 and 1931 than parishes with burghs or close to burghs. In this period, only two burghs – Elgin and Inverness – displayed rises in population. This pattern of change continued through to the 1951 census. The dramatic difference in population change between the eastern and western parishes in Sutherland and Ross over the first half of the century was now clear: within the latter county, the population fell by 25 per cent in four western parishes but rose by 11 per cent in four parishes in the east between 1931 and 1951, and a similar though less marked pattern was evident in Sutherland.[4]

Population in each of the counties around the Moray Firth from 1911 to 1951.

|  | 1911 | 1921 | 1931 | 1951 |
|---|---|---|---|---|
| Banff | 61,402 | 57,298 | 54,907 | 50,148 |
| Caithness | 32,010 | 28,285 | 25,656 | 22,710 |
| Inverness | 87,272 | 82,455 | 82,108 | 84,930 |
| Moray | 43,427 | 41,558 | 40,806 | 48,218 |
| Nairn | 9,319 | 8,790 | 8,294 | 8,719 |
| Ross and Cromarty | 77,364 | 70,818 | 62,799 | 60,508 |
| Sutherland | 20,179 | 17,802 | 16,101 | 13,670 |

Blockhouses on the South Sutor mark where guns were sited to defend the entrance of the Cromarty Firth during the First World War.

The Second World War from its outbreak in September 1939 until it ended in 1945 had a significantly different impact on the region in that much more of the military activity took place 'at home'. The Royal Air Force established a considerable presence around the Firth with a series of airfields. Some were involved in the defence of the naval bases in Scapa Flow and the Cromarty Firth, whereas others were home to squadrons charged with patrolling northern waters, training or operating against Nazi-occupied Norway. Dalcross and Wick were among the fields that became civilian airports after the War, while Kinloss and Lossiemouth remained as RAF stations until 2011. Especially in the early months, Wick, Fraserburgh and Peterhead suffered air raids and loss of civilian life, with Fraserburgh being nicknamed 'Little London' for the damage it sustained. As in the First World War, restrictions were placed on travel with all the country north of Inverness once again a protected area. Fear of invasion led to the formation of the Home Guard and to the building of coastal defences wherever it seemed likely troops and tanks could attempt to land. Many of these installations – pillboxes, dragons' teeth to impede tanks and so on – remain half hidden in the dunes of Culbin and Sinclair's Bay to this day, as do the gun emplacements on the Sutors of Cromarty. Refugees and troops from some of our occupied neighbours made their way to our shores, Poles and Norwegians prominent among them and, in localities with woodland, contingents of foresters from Canada, Newfoundland and, for a time in Golspie, from British Honduras (Belize). A severe blow was struck to local morale in June 1940 when the 51st Highland Division, still fighting in France after most of the British troops had been evacuated through Dunkirk, was surrounded and forced to surrender at the little Normandy town of Saint Valery-en-Caux. The Division was reconstituted and fought in North Africa, Italy and northern Europe. One community, Inver on the Dornoch Firth, had a unique experience: in November 1943 the inhabitants were told to abandon their homes within the month. Much of the peninsula of Tarbat Ness was taken over for secret rehearsals for the D-Day landings although in the event the beaches at Inver proved to be no use for training, and the people were allowed back to their homes in the summer of 1944. The War added a fresh list of names to the memorials erected in the 1920s in memory of the fallen in 1914–18, and also sealed in place many changes that were already under way before it began but accelerated after 1945, among them the great shift from steam power to oil fuel, the displacement of the horse by the tractor as the main motive power

in working the land, and the final fading of the drift netting for the herring in favour of new methods.

The very recent past, the latter half of the twentieth century, is still too close, in a sense too raw, to allow the easy discernment of significant trends. The picture has to be painted with a broad brush. The immediate post-War years saw the introduction of the welfare state, one aspect of a much greater deployment of technocratic planning in almost every aspect of life. The Second World War itself produced the sense that now there was an opportunity for a new start, for doing things better, but it took years for this 'progress' to become manifest. A major effort to boost economic development in the Highlands came from the foundation of the Highlands and Islands Development Board (HIDB) in 1965 with its headquarters in Inverness. Its region of responsibility was, as its name suggests, initially the seven 'crofting counties' but, with the reorganisation of local government in 1975, this was expanded to take in Nairn and the upland part of Morayshire, around Grantown-on-Spey. As if in admission of a hitherto absurdly rigid distinction between Highland and Lowland, the HIDB was to expand further and eventually its successor, Highlands and Islands Enterprise,

Kilmorack power station on the lower Beauly river, constructed in the early 1960s for the North of Scotland Hydro-Electric Board, brought new hope for the economic development of the region after the Second World War.

211

Low tide in the Cromarty Firth sees common seals and cormorants making good use of exposed rocks.

founded in 1990, took into its jurisdiction all of Moray. One of the first bodies of its kind in Europe, the HIDB was founded on the premise that an agency of government, dedicated to development, could kick-start a new economic future for a region. In the opening pages of its first annual report, under the heading 'The challenge', the HIDB identified depopulation as the central problem and attacked those, usually outsiders, who saw the Highlands as inhabited by 'crofters of an idealised character exempt from the ordinary laws of making a real livelihood' and wanted to preserve it as a 'natural relief valve for an over-urbanised country'.[5] The HIDB was equally determined to resist the desires of those who wanted to see sweeping land reform wipe away the great estates. Three main 'props' were named as key elements in the economy: the primary producing activities of farming, forestry and fishing, tourism and manufacturing industry. In the first four years of its existence, the HIDB invested over £100,000 in projects and schemes. A pocket of industrialisation was promoted on the north shore of the Cromarty Firth, possibly influenced by the apparent success of the establishment of the Dounreay nuclear reactor project in Caithness in reinvigorating economic life on the north coast. Europe's largest whisky grain distillery had opened in Invergordon in 1960, and there were also plans to develop a petrochemical plant. An aluminium smelter opened in Invergordon in 1971. The coincidental discovery of large oil and gas reserves in the North Sea in the late 1960s was a bonus, spurring the establishment of yards to build offshore equipment and production platforms at Nigg and Ardersier, and an oil terminal also at Nigg, in the lee of the North Sutor. The demand for labour far outstripped the local resources and precipitated an influx of workers from other parts of the country. Alness, Invergordon, Nairn and Culloden and their neighbours grew as boom towns, leaving the locals somewhat taken aback by the abrupt change in the world they knew. The HIDB could report with some satisfaction in 1972 that the population in the area had risen by 5,000 since 1966; such an increase had previously taken the 45 years since 1921. The industrial phase did not last long. The petrochemical plant remained a dream, and the smelter ceased to produce gleaming ingots of aluminium in 1981, closing its doors with shocking abruptness just before Christmas and laying off almost 900 workers in one stroke. The construction yards at Nigg and Ardersier experienced fluctuating fortunes before their closures. The boom gave way to high levels of unemployment and social problems, still not fully resolved, and there arose a tacit acceptance that placing so much dependence on single industrial enterprises was not a wise thing to do. By the 1980s the eyes of the developers were falling back on the traditional and the small scale.

The Second World War put an end to the steamers which used to link Wick with Aberdeen and Leith and for a while it looked as if much of the rail services would also be slashed. The infamous Beeching axe, named after the man charged with making British Rail profitable, suggested cutting the lines from Inverness to Caithness, and all the branch lines in the north-east, leaving only direct, non-stopping trains running between Inverness and Aberdeen. After a great campaign under the catchy heading of MacPuff, the main lines stayed open. Air passenger services across the Firth began in 1933 when Captain Ernest 'Ted' Fresson, a former Royal Flying Corps pilot and barnstormer with an air circus, and Robert Donald, director of the Inverness motor company Macrae and Dick, formed Highland Airways Ltd and introduced regular flights between Wick and Inverness. In 1934 the flying network expanded to include Aberdeen and, at almost the same time another company, Allied Airways, also began to connect Aberdeen with the north, provoking an intense rivalry in the Firth skies until the government settled a division of routes.

Looking west up the length of the Cromarty Firth towards the hills of Ross and Inverness-shire. These fjord-like inlets show how sea transport could thrive before modern road and rail.

The most important legacy of the industrial phase is the improved road transport in the inner Moray Firth coastlands. The three firths – the Cromarty, the Inverness and the Dornoch were bridged in 1980, 1982 and 1992 respectively. These crossings dramatically cut the distances between the Firth communities. With other road improvements and despite the occasional unresolved choke point, such as at Berriedale, the bridges halved the time it took to drive between Caithness and Inverness. A commuting zone sprang up within an hour's drive of the Highland capital. This brought about a revolution in shopping habits, to the benefit of the retail trade in Inverness but to the detriment of shopkeepers in the smaller towns. Such was the centralising effect of the Invergordon–Inverness growth zone that talk began of a linear city emerging along the A9. In contrast to the made-over north road, the main artery linking Inverness, Nairn and Moray to Aberdeen, the A96, remained and remains a poor cousin.

In the winter of 1966–67 great shoals of herring invaded the Inverness and Beauly Firths, boosting the customary Kessock herring fishery to an

almost unheard-of level. Local and foreign fishermen, including Klondikers from Germany, Norway and Iceland, cast their nets in the crowded sea and the *Inverness Courier* recorded that 3,000 cran was an average daily catch. Towards the end of February it was estimated that around 135,000 cran of fish had been taken since the season had started in October. It was a dramatic but short-lived blip in a downward trend. Around 1950 the seine net boats working from Wick fished around an hour or two's steaming from the port, relying on hard-won knowledge and seamarks as their guides. It was common to spend only four or five days at sea, if that, before heading homeward with a catch. The smaller boats in the inshore fleets, setting creels for crab and lobster, also had their fishing effort curtailed by their gear and the seasons. In Keiss, for example, close to the northern corner of the Firth, the crab fishermen worked from around Easter time until November, and sought other jobs in winter when, anyway, the crabs migrated offshore, beyond the safe limits of their boats. It was still possible then to try line fishing in January and February and hook impressive cod, skate and halibut within sight of the shore. Although not many people may have realised it at

Late sun bathes the curve of the Cromarty Bridge, the first of the major infrastructure projects of the 1980s.

215

Buckie and its harbour.

the time, the future for fishing in the Moray Firth in the 1950s was bleak. Looking back, it is possible, although it would be difficult to confirm, that it was during this decade that the fish stocks in the Firth itself became exploited past their sustainable maximum. Vessels working from the smaller ports such as Brora, which was home to two seine net boats in 1955, Burghead, Nairn and Inver failed to make a living as their stocks collapsed under the fishing pressure from bigger neighbours. In 1953 Helmsdale had 11 seiners, in 1984 there were only two. The Wick fishing fleet boasted 47 registered boats in 1952, mostly seine netters and, remarkably, a single sailing vessel. The herring fishing at that time was in steep decline and soon ceased altogether. By 1982 the Wick fleet had fallen to 18 seiners. The decline was masked for many years by several eventualities. Enterprising fishermen in Lossiemouth and the neighbouring harbours began to rove far in search of catches, to the west coast of Scotland and down to Ireland, out into the

North Sea and up around Orkney, and for them the life was adventurous and rewarding. Bigger boats were built, electronic technology pinpointed shoals with accuracy, synthetic materials lengthened the working lives of the gear. The hydraulic power block enabled the handling of the purse seine, designed to surround and capture whole shoals of pelagic mackerel and herring. From the 1950s, prawns and squid became significant catches in their own right; until then they had been largely ignored in favour of white fish. Dredging for clams started, with unknown long-term consequences for the seabed, and even more unlikely fisheries sprang into being, for example for the buckie and the spoot [razor clam], as the struggle to win a living from the sea diversified and as membership of the European Union opened up new markets. A severe economic crisis, with rising costs, falling prices and the loss of men to the better-paying oil industry, hit the fishing fleets in 1974, forcing the fishermen to lay aside their strong sense of individuality

Buckie harbour.

217

The harbour at Banff now shelters a marina as the towns around the Firth find new roles for old facilities.

and resort to blockading ports to protest their grievances. The introduction of the Common Fisheries Policy in the 1970s brought in its wake a new spate of adjustments to be made in a strongly traditional way of life.

The 1980s were marked by a significant growth in environmental awareness among the general public. History and development began to be viewed in 'green' terms, as what was sustainable and what was unacceptable exploitation became the focus of sharp debate. People have come to see the economic value in 'ecosystem services', the benefits conferred by the environment that have hitherto been taken for granted. The Moray Firth as a region where there is still a relatively small population dispersed through attractive and rich natural surroundings appears to offer the opportunity to achieve a desirable balance between development and conservation. This kind of thinking has become a shaping influence among the planners and the politicians. Highlands and Islands Enterprise, the economic and community development agency responsible now for all the Firth coast except the Banff section in Aberdeenshire, has announced six topics on which it will concentrate its efforts – energy, life sciences, food and drink, the creative industries, tourism, and financial and business services.[6] The

energy focus is on generating power from renewable sources, in keeping with the principle of sustainability, and the two production platforms on the only oilfield in the Firth, the Beatrice field some 12 miles off the Caithness coast, are set to be joined by the turbines of two great wind farms in the near future. The population of bottlenose dolphins in the Firth, the most northerly members of this species in the world's oceans, has come to symbolise the natural history of the region and supports several wildlife tour businesses sailing from the quays where the herring fleets used to land their catches. The marina is now more common than the fish mart. Tourism matters hugely to the economy, and as a result every place tries to find a unique attraction to give it an identity in the brochures and tourist websites, so that it seems there is hardly a historic site – ruined castle, or tumbledown Neolithic cairn, or old kirk – without its interpretive signboard. The locals benefit from this in many cases possibly more than visitors who tend to be guided around a standard series of oases, and now there is a wide interest in local history among residents. The most obvious example of the promise of the so-called creative and intellectual industries is the establishment of the University of the Highlands and Islands, a collegiate institution with a scatter of campuses across the north.

A wind farm beside the Causewaymire on the Caithness moors.

219

Inverness and its sprawling suburbs is the largest settlement in the Firth region. Here both the River Ness and the Caledonian Canal open into the Firth, and the Kessock Bridge carries the A9 over the tide races. On the conifer-clad ridge of Craig Phadraig beyond the city are the vitrified remains of an Iron Age fort.

The land itself remains the lasting source of wealth, and the processing of the grain and the meat from the region's fields is now by far the most important sector of manufacturing. Whisky is the prime example of an ancient art that has become big business. Blessed with abundant barley, peat and clean water, the glen of the Livet river became a centre of excellence for malt whisky, with some 200 small stills in the 1820s, and is now at the centre of the Strathspey whisky industry. The road signs to Moray remind arrivals they are entering 'Malt Whisky Country', and there are many famous distilleries at other locations around the Firth in Wick, Brora, Tain, Muir of Ord and several other places.

The census in 2001 found that there were 88,940 people in Moray, 35,171 in Banff and Buchan, and 208,914 in Highland, two-thirds of whom lived in local authority wards fringing the Moray Firth – more than a quarter of a million souls, as Alexander Webster may have phrased it in 1755. Most

A modern trawler on the slip at Macduff Shipyard. The yard has been turning out steel and wooden vessels since 1940.

of them, however, now live in the town rather than the countryside. The 2001 census counted, for example, 42,880 in Inverness, the largest urban centre, and 20,929 in Elgin. The present dispensation for local government has whipped the old counties from official maps but they live on in the minds of the people. The attentive ear can tell from a shift of accent or a change in dialect when an old boundary has been crossed. There is no doubt that over the last century and accelerating in the last 50 years, improving communications have resulted in an erosion of the local, the little things that mark a place as unique. Instead there has evolved a more all-embracing 'way of life', sped on by mass media and the movement of workers. At the same time, a regional consciousness has tentatively been appearing, an increasing awareness of common interests that span the Firth. We who live around the Firth all dip our toes in the same water, so to speak. This notion informs the work and ambitions of the Moray Firth Partnership, whose website states, 'We help people and organisations find new ways to communicate and work together to keep the Moray Firth's natural, economic and social resources in good heart, now and for future generations'.[7] The future will probably see this trend go further. There is already talk of local authorities

The pier at Burghead points towards an unsettled sky over the inner Firth.

sharing services to reduce costs and if services are shared can mergers of councils be so far behind? Are we returning to a commonality in life around the Firth, something that was perhaps known to the Picts and something that has probably existed for a long time among seafarers? The divisions and conflicts associated with the clash of cultures that first spurred me to tackle this book persist only as a fading echo. The history of dynasties, warlords and tribes has given way to a more all-encompassing collective history.

Opposite, top: Chanonry Point and its lighthouse is a favourite spot for visitors hoping to catch sight of the Moray Firth dolphins. Bottom: A trio of dolphins leaping off the coast. Wildlife tourism to see these animals and other species around the Firth now contributes substantial sums to the local economy.

# NOTES AND SOURCES

## Introduction

1. Graham, 1977.
2. Mowat, 2003.
3. Worthington, 2011.
4. Shaw, 1882.
5. *Statistical Account of Scotland* (OSA), 1793, *V*, Elgin, p. 5.

## 1 Long Reach of Empire

1. Tacitus, 1970. The location of Mons Graupius has never been established. A popular choice is the hill of Bennachie in Aberdeenshire, but a plausible case has been presented recently by James E. Fraser for it being further south, in the vicinity of the Cairnie Braes in Strathearn (Fraser, 2005).
2. Comber, 1995.
3. See, for example, Strang, 1998, and Breeze, D.J., 1990.
4. Possible and confirmed Roman sites are listed on the websites of the Historic Environment Record: http://her.highland.gov.uk and http://www.aberdeenshire.gov.uk/smrpub. See also http://canmore.rcahms.gov.uk
5. Dodghson, 1980.
6. McIntyre, 1998.

## 2 Emergence of a Pictish Nation

1. Hunter, 2007.
2. Ritchie, 1976.
3. Small, 1983.
4. Historic Environment Record: http://her.highland.gov.uk. ID MHG17924. For details of the research at Butser, see Reynolds, 1995.
5. Hunter, 2002.
6. *De Situ Albanie*, quoted in Anderson, 1990 [1922], Vol. 1, p. cxvii.
7. Woolf, 2006.
8. Shaw, 1882, p. 27.
9. Quoted in Anderson, 1990 [1922], Vol. 1, p. 43.

10. Adamnan, 1988.
11. Fraser, 2009, p. 100.
12. Allan, 2005, p. 14.
13. Anderson, 1990 [1922], Vol. 2, p. 174.
14. Laing, 2000, recognises four classes of symbol stone, whereas Small, 1983, has three, and Carver, 1999, only two.
15. Edwards and Ralston, 1977–78.
16. Woolf, 2006.

## 3 Invaders

1. Quoted in Laing, 1994.
2. Watson, 1993 [1926].
3. For details of the complex politics of the period, see Woolf, 2007.
4. All the quotations from the various annals are from Anderson, 1990 [1922].
5. Anderson, 1990 [1922], Vol. 1, p. 259.
6. Lamb, 1977, Vol. 2, p. 371.
7. *Heimskringla*, 1961, p. 65.
8. Anderson, 1990 [1922], Vol. 1, p. 331.
9. Research on DNA in northern populations is still going on. The picture that has emerged so far supports the historical pattern of Norse settlement in the far north. Nearly one half of Orkney paternal lineages are Norse in origin. The figures for the Western Isles and Caithness are around one-third and one-quarter respectively. About 8 per cent of Caithness lineages are Irish in origin, reflecting the northward movement of Gaels. The majority of the population, however, has ancestry dating back to the indigenous inhabitants in prehistoric times. See, for example, Oppenheimer 2007, and Moffat and Wilson, 2011.
10. Anderson, 1990 [1922], Vol. 1, p. 273.
11. Watson, 1904.
12. *Orkneyinga Saga*, 1981, p. 27.

## 4 Forging of a Province

1. Anderson, 1990 [1922], Vol. 1, p. 270.
2. Whittington, 1974–5.
3. Nicolaisen, 1976, p. 154.
4. Book of Deer quoted in Anderson, 1990 [1922], Vol. 2, p. 174.
5. In a detailed study of place names in the Beauly area, Simon Taylor has identified several names that are possibly Pictish in origin – Altyre, Dunballoch, Erchless, Groam and Urchany. See Taylor, 2002.
6. For example, in the introduction to *Registrum Episcopatus Moraviensis*, 1837. Watson thought 'Forn' was possibly a misreading for 'Forir', i.e. the Varar.
7. Nicolaisen, 1993.
8. Anderson, 1990 [1922], Vol. 1, p. 271.
9. Anderson, 1990 [1922], Vol. 1, p. 271.

10. Taylor, 2002, mentions Eskadale, Plodda and Ruttle as Norse place names in Strathglass. This area could have been the locus of trade and possibly settlement by Norse in pursuit of good timber.
11. Bower, 1996, Vol. 2, p. 349.
12. Anderson, 1990 [1922], Vol. 1, p. 576.
13. The quote is from the writings of an Irish monk and is given in Barrow, 1981, p. 26.

## 5 A New Order

1. Anderson, 1990 [1922], Vol. 2, p. 174.
2. Anderson, 1990 [1922], Vol. 2, p. 174.
3. Tayler, 1937.
4. Anderson, 1990 [1922], Vol. 2, p. 302.
5. Anderson, 1990 [1922], Vol. 2, p. 312.
6. *Orkneyinga Saga*, 1981, p. 221.
7. Anderson, 1990 [1922], Vol. 2, p. 347.
8. Young, 1997.
9. The Charlemagne story is given in Drummond-Norie, 1898.
10. Anderson, 1990 [1922], Vol. 2, p. 471.
11. Anderson, 1990 [1922], Vol. 2, p. 404.

## 6 Burghs and Fiefdoms

1. Duncan, 1978, p. 475.
2. Coleman, 2004.
3. Douglas, 1934, p. 15.
4. Munro, 1984, p. 133.
5. OSA XII, 1794, Nairn, p. 383.
6. Bain, 1881, p. 446.
7. Bain, 1884, p. 436.
8. Barrow, 1971, p. 223.
9. See Anderson, 1967, and McNeill and Nicholson, 1975, Map 31.
10. Anderson, 1977, p. xlvi.

## 7 Occupation

1. Bain, 1881.
2. Quoted in Barron, 1934.
3. McAndrew, 1999.
4. Bain, 1884, p. 300.
5. Bain, 1884, p. 434.
6. Barrow, 1976, p. 245.
7. Barbour, J., *The Bruce*, Book IX, lines 301–305 (1909 edition, p. 156).

## 8 Highland and Lowland

1. Miller, 2004, p. 50.
2. Douglas, 1934, p. 34.
3. Miller, 2004, p. 94.
4. The Bremen cog is on display in the maritime museum in Bremerhaven.
5. Cramond, 1903, Vol. 1, p. 19.
6. Lynch, 2001, p. 487.
7. Lamb, 1977, Vol. 2, p. 451.
8. Grant, 1996, p. 61.
9. OSA, 1794, XI, p. 138.
10. OSA, 1795, XIV, p. 400.
11. Hume Brown, 1893, p. 10.
12. Martin, 1994, p. 207.
13. Burt, 1998, p. 204.
14. Hume Brown, 1893, p. 11.
15. Johnston, 1997.
16. Mackintosh, 1903; and Paton, 1903.
17. Mackenzie, 1898.
18. Mackenzie, 1894.
19. Mackenzie, 1891.
20. Calder, 1887.
21. Grant, 1993.
22. Quoted in Cramond, 1903, Vol. 1, p. 18.
23. Fraser, 1905, p. 95.

## 9 Struggles in Kirk and State

1. Quoted in Hume Brown, 1893, p. 74. 'Crukis in with ane gret discens and vale' describes the land bending inwards to form 'a great slope and valley'. Boece wrote the original in Latin, the Scots translation was by John Bellenden in 1536.
2. Quoted in Donaldson, 1970, p. 102.
3. Cramond, 1891, Vol. 2, p. 10.
4. MacGill, 1909, p. 7.
5. Bain, 1928, p. 172.
6. Craven, 1908, p. 25.
7. *Fasti Ecclesiae Scoticanae*, 1926, Vol. 6, p. 298.
8. In this verse of the ballad, the composer has made the king speak.
9. *Register of the Privy Council of Scotland*, II, p. 148.
10. Author's translation of Scots original. In 'The Chronicle of Aberdeen', 1842, p. 47.
11. Larner, 1981.
12. Miller, 1999, p. 18. The author also heard of a crofter who used to perform a ritual perambulation of the field with a new, empty pail before putting the cow out to graze for the first time after winter, presumably to ensure a good milk yield.

13. Quoted in Lippe, 1890, p. 272.
14. Tayler, 1980.
15. Spalding, 1850, Vol. 1, p. 3.
16. Spalding, 1850, Vol. 1, p. 60.
17. Spalding, 1850, Vol. 1, p. 50.
18. James Gordon, *History of Scots Affairs*, quoted in Cramond, 1891, Vol. 1, p. 94.
19. Spalding, 1850, Vol. 1, p. 135.
20. Spalding, 1850, Vol. 2, p. 447.
21. Fraser, 1905, p. 295.
22. Spalding, 1850, Vol. 2, p. 473.
23. Fraser, 1905, p. 296.
24. Fraser, 1905, p. 313.
25. Fraser, 1905, p. 315.
26. Cramond, 1903, Vol. 2, p. 356.
27. Fraser, 1905, p. 396.
28. Cramond, 1903, Vol. 2, p. 279.
29. Douglas, 1934, p. 52.
30. Fraser, 1905, p. 415.
31. Fraser, 1905, p. 412.
32. Fraser, 1905, p. 420.
33. Cullen, Whatley and Young, 2006.
34. Mitchell, 1902.
35. Dr K.J. Cullen, presentation, 'Famine in the Late 17th Century', University of the Highlands and Islands Centre for History, Dornoch, 24 April 2010.
36. The Jacobite risings have been documented and analysed many times. See, for example, Paton, 1895; Prebble, 1962; Livingstone et al., 2001; Duffy, 2007; and for events specifically in Inverness, Miller, 2004.

## 10 Changing the Face of the Land

1. Tayler, 1914, Vol. 1, p. 63.
2. The figure of £800,000 is given in Millar, 1909, p. xxxiii. There are various accounts of the ill-fated Darien scheme to create a Scots colony in Panama; see for example Watt, 2007.
3. Highland Archive, Inverness. CTI/BI 24/1.
4. Donaldson, 1984.
5. OSA, IX, 1793, Birnie, p. 161.
6. OSA, V, 1793, Elgin, p. 6.
7. OSA, XIX, 1797, Boyndie, p. 306.
8. OSA, IX, 1793, Lhanbryde, p. 174.
9. OSA, VIII, 1793, Cromdale, p. 255.
10. OSA, X, 1793, Wick, p. 26.
11. Millar, 1909, p. xii.
12. Adams, 1979, p. xxi.
13. OSA, XX, 1798, Banff, p. 326.
14. OSA, XIX, 1797, Auldearn, p. 623. The boll was a measure of volume for dry

goods, equal approximately to 145 litres. A boll of meal was approximately 63.5 kg, or 140 lb.

15. OSA, XII, 1794, Kirkmichael, p. 426.
16. OSA, III, 1790, Deskford, p. 359.
17. OSA, X, 1793, Wick, p. 22. The classic work on drove roads is Haldane, 1973. See also Baldwin, 1986.
18. OSA, V, 1793, Keith, p. 420.
19. OSA, XI, 1794, Huntly, p. 467.
20. OSA, IV, 1792, Ardclach, p. 153.
21. Mitchell, *Geographical Collections*, Vol. 1, p. 39.
22. There are many dialect names for animals and plants in the communities around the Firth. Those mentioned by Hepburn and Garden include seath – coley, podler – young coley, parten – crab, scath – probably cormorant, sea coulter – puffin, taster – guillemot, maw – gull, whap – curlew.
23. Mitchell, 1906, Vol. 2, p. 269.
24. OSA, XIV, 1795, Speymouth, p. 394.
25. OSA, X, 1793, Wick, p. 27.
26. OSA, IV, 1792, Drainy, p. 79; and OSA, V, 1793, Elgin, p. 6.
27. Mitchell, 1906, Vol. 1, p. 459.
28. OSA, IV, 1792, Ardclach, p. 151.
29. OSA, XX, 1798, Dyke, p. 226.
30. OSA, V, 1793, Elgin, p. 10.
31. OSA, XII, 1794, Kirkmichael, p. 470.
32. OSA, X, 1793, Clyne, p. 302.
33. OSA, X, 1793, Wick, p. 20.
34. OSA, III, 1792, Dingwall, p. 3.
35. OSA, VIII, 1793, Cromdale, p. 253.
36. OSA, XVII, 1796, Mortlach, p. 429.
37. OSA, XVIII, 1796, Botriphnie, p. 644.
38. OSA, XVII, 1796, Mortlach, p. 426.
39. OSA, IV, 1792, Cawdor, p. 355.
40. OSA, VIII, 1793, Moy and Dalarossie, p. 499.
41. OSA, XII, 1794, Cromarty, p. 254.
42. OSA, XIX, 1797, Alness, p. 236.
43. Grant, 1972, p. 3.
44. Miller, 1869 [1834], p. 473.
45. Miller, 1869 [1834], p. 477.
46. OSA, VIII, 1791, Canisbay, p. 145.


*11 Sheep, Roads and Herring*

1. OSA, VIII, 1791, Dornoch, p. 5.
2. This list is compiled from Richards, 2008 [1982].
3. OSA, XIX, 1797, Alness, p. 235.
4. OSA, XX, 1798, p.xiii.
5. OSA, VIII, 1791, p. 376.
6. Bangor-Jones, 2002.

7. Mitchell, 1884.
8. Macleod, 1892, p. 9.
9. Richards, 1973.
10. Millward, 1964, p. 221.
11. Thomas, 1977.
12. See, for example, Devine, 2003, and http://www.spanglefish.com/
slavesandhighlanders.
13. *New Statistical Account of Scotland* (NSA), XIV, 1845, p. 66.
14. NSA, XV, 1845, p. 93.
15. NSA, XV, 1845, p. 41.
16. The great herring fishery has been extensively studied: see, for example,
Dunlop, 1978; Sutherland, 1984; Miller, 1999; Coull, 2008.
17. Sage, 1899, p. 109.
18. NSA, XIII, 1845, p. 42.
19. NSA, XIV, 1845, p. 359.
20. NSA, XV, 1845, p. 157.
21. NSA, XIII, 1845, p. 78.
22. NSA, XIII, 1845, p. 91.
23. NSA, XIII, 1845, p. 108.
24. NSA, XIV, 1845, p. 279.
25. NSA, XIV, 1845, p. 28.
26. NSA, XIV, 1845, p. 239.
27. NSA, XV, 1845, p. 162.
28. NSA, XV, 1845, p. 203.
29. *Book of Banff*, 2008, p. 66.
30. *Inverness Courier*, 23 May 1832.
31. Brown, 1893, p. 21.
32. Lauder, 1998 [1873], p. 236.
33. *Inverness Courier*, 1 August 1832.
34. Coull, 2008, p. 214.
35. North Highland Archive, Wick, Wick Harbour Master's Logbook WHT/
Ga/1.

## 12 Roots of the Future

1. *People's Journal*, 13 January 1900.
2. *Highland News*, 24 April 1926.
3. *Inverness Courier*, 18 May 1926.
4. Kyd, 1952.
5. Highlands and Islands Development Board, 1st Annual Report, 1967, p. 2.
6. Highlands and Islands Enterprise, Operating Plan 2011–14, available on
http://www.hie.co.uk.
7. Moray Firth Partnership, http://www.morayfirth-partnership.org.

# BIBLIOGRAPHY

Books and other sources with information on the Moray Firth and its history abound; those listed below are a starting point, and contain many references to further works.

Adam, R.J. (ed.) (1991) *The Calendar of Fearn: Text and Additions 1471–1667* (Edinburgh, Scottish Historical Society).

Adamnan (1988) *Life of Columba*, ed. W. Reeves (Lampeter, Llanerch Enterprises reprint).

Adams, I.H. (ed.) (1979) *Papers on Peter May, Land Surveyor, 1749–1793* (Edinburgh, Scottish Historical Society).

Allan, N. (2005) *The Celtic Heritage of the County of Banff* (Elgin, Moravian).

Alston, D. (2006) *My Little Town of Cromarty* (Edinburgh, Birlinn).

Anderson, A.O. (1990 [1922]) *Early Sources of Scottish History AD 500 to 1286*, new edition, ed. M. Anderson (Stamford, Watkins).

Anderson, J. (ed.) (1977) *The Orkneyinga Saga*, trans. J.A. Hjaltalin and G. Goudie, facsimile edition (Edinburgh, Mercat).

Anderson, M.L. (1967) *A History of Scottish Forestry* (London, Nelson).

Anson, P.F. (1930) *Fishing Boats and Fisher Folk on the East Coast of Scotland* (London, Dent).

Ash, M. (1991) *This Noble Harbour: A History of the Cromarty Firth* (Edinburgh, John Donald).

Atherton, M. (1992) *Upland Britain: A Natural History* (Manchester: Manchester University Press).

Bain, G. (1928) *History of Nairnshire* (Nairn).

Bain, J. (ed.) (1881) *Calendar of Documents Relating to Scotland Preserved in Her Majesty's Public Record Office, London*, Vol. 1 (Edinburgh).

Bain, J. (ed.) (1884) *Calendar of Documents Relating to Scotland Preserved in Her Majesty's Public Record Office, London*, Vol. 2 (Edinburgh).

Baldwin, J.R. (ed.) (1982) *Caithness: A Cultural Crossroads* (Edinburgh, Scottish Society for Northern Studies).

Baldwin, J.R. (1986) 'The Long Trek: Agricultural Change and the Great Northern Drove', in J.R. Baldwin (ed.) *Firthlands of Ross and Sutherland* (Edinburgh, Scottish Society for Northern Studies).

Bangor-Jones, M. (1986) 'Land Assessments and Settlement History in Sutherland and Easter Ross', in Baldwin, J.R. (ed.) *Firthlands of Ross and Sutherland* (Edinburgh, Scottish Society for Northern Studies).

Bangor-Jones, M. (2002) 'Sheep Farming in Sutherland in the Eighteenth Century', *Agricultural History Review* 50(2), 181–202.

Barbour, J. (1909) *The Bruce*, ed. W.M. Mackenzie (London, A. & C. Black).

Barrett, J.H. (1997) 'Fish Trade in Norse Orkney and Caithness: A Zooarchaeological Approach', *Antiquity* 71, 616–638.

Barron, E.M. (1934) *The Scottish War of Independence: A Critical Study* (Inverness).

Barrow, G.W.S. (ed.) (1960) *The Acts of Malcolm IV* (Edinburgh, Edinburgh University Press).

Barrow, G.W.S. (ed.) (1971) *The Acts of William I* (Edinburgh, Edinburgh University Press).

Barrow, G.W.S. (1976) *Robert Bruce and the Community of the Realm of Scotland* (Edinburgh, Edinburgh University Press).

Barrow, G.W.S. (1980) *The Anglo-Norman Era in Scottish History* (Oxford, Oxford University Press).

Barrow, G.W.S. (1981) *Kingship and Unity: Scotland 1000–1306* (London, Arnold).

Batey, C.E., Jesch, J. and Morris, C.D. (eds) (1993) *The Viking Age in Caithness, Orkney and the North Atlantic* (Edinburgh, Edinburgh University Press).

Beaton, E. (1995) *Sutherland: An Illustrated Architectural Guide* (Edinburgh, RIAS/Rutland Press).

Beaton, E. (1996) *Caithness: An Illustrated Architectural Guide* (Edinburgh, RIAS/Rutland Press).

Black, G.F. (1993) *The Surnames of Scotland* (Edinburgh, Birlinn).

*Book of Banff* (2008) (Halsgrove, Banff Preservation and Heritage Society).

Bower, W. (1996) *Scotichronicon*, ed. D.E.R. Watt (Aberdeen, Aberdeen University Press).

Breeze, D.J. (1990) 'Agricola in the Highlands?', *Proceedings of the Society of Antiquaries of Scotland* 120, 55–60.

Breeze, D.J. (2008) *Edge of Empire: The Antonine Wall* (Edinburgh, Birlinn).

Brown, K.M. (1986) *Bloodfeud in Scotland 1573–1625* (Edinburgh, John Donald).

Brown, T. (1893) *Annals of the Disruption* (Edinburgh, Macniven & Wallace).

Bulloch, J.M. (ed.) (1903) *The House of Gordon* (Aberdeen, New Spalding Club).

Burt, E. (1998) *Letters from the North of Scotland* (Edinburgh, Birlinn).

Calder, J.T. (1887) *Sketch of the Civil and Traditional History of Caithness* (Wick).

Cameron, D.K. (1978) *The Ballad and the Plough* (London, Gollancz).

Cameron, D.K. (1980) *Willie Gavin, Crofter Man* (London, Gollancz).

Cameron, D.K. (1986) *The Cornkister Days* (London, Gollancz).

Carver, M. (1999) *Surviving in Symbols: A Visit to the Pictish Nation* (Edinburgh, Birlinn).

'Chronicle of Aberdeen' (1842) *The Miscellany of the Spalding Club*, Vol. 2 (Aberdeen), pp. 29–70.

Clark, G. (2009) *Redcastle: A Place in Scotland's History* (London, Athena).

Clough, M. (1990) *Two Houses* (Aberdeen, Aberdeen University Press).

Coleman, R. (2004) 'The Archaeology of Burgage Plots in Scottish Medieval Towns: A Review', *Proceedings of the Society of Antiquaries of Scotland* 134, 281–328.

Comber, D.P.M. (1995) 'Culbin Sands and the Bar', *Scottish Geographical Magazine* 111 (1), 54–57.

Coull, J. (2008) 'The Herring Fishery' in J.R. Coull, A. Fenton and K. Veitch (eds) *Scottish Life and Society: Boats, Fishing and the Sea* (Edinburgh, John Donald).

Cowley, D.C. and Stevenson, J.B. (2001) *The Historic Landscape of the Cairngorms* (Edinburgh, RCAHMS).

Cramond, W. (ed.) (1891) *The Annals of Banff*, 2 vols (Aberdeen, New Spalding Club).

Cramond, W. (ed.) (1903) *The Records of Elgin* (Aberdeen, New Spalding Club).

Craven, J.B. (1908) *A History of the Episcopal Church in the Diocese of Caithness* (Kirkwall, Peace).

Crawford, B.E. (1982) 'Scots and Scandinavians in Medieval Caithness: A Study of the Period 1266–1375', in J.R. Baldwin (ed.) *Caithness: A Cultural Crossroads* (Edinburgh, Scottish Society for Northern Studies).

Crawford, B.E. (1985) 'The Earldom of Caithness and the Kingdom of Scotland 1150–1266', in K.J. Stringer (ed.) *Essays on the Nobility of Medieval Scotland* (Edinburgh, John Donald).

Cullen, K.J., Whatley, C.A. and Young, M. (2006) 'King William's Ill Years: New Evidence on the Impact of Scarcity and Harvest Failure during the Crisis of the 1690s on Tayside', *Scottish Historical Review* 85(2), 253–280.

Cunliffe, B. (2001) *Facing the Ocean: The Atlantic and Its Peoples* (Oxford, Oxford University Press).

Cunningham, I.C. (ed.) (2001) *The Nation Survey'd: Timothy Pont's Maps of Scotland* (East Linton, Tuckwell).

Dark, P. (2000) *The Environment of Britain in the First Millennium* (London, Duckworth).

Devine, T.M. (1993) *The Great Highland Famine* (Edinburgh, John Donald).

Devine, T.M. (2003) *Scotland's Empire 1600–1815* (London, Allen Lane).

Dodghson, R.A. (1980) 'Medieval Settlement and Colonisation', in: Parry, M.L., Slater, T.R. (eds), *The Making of the Scottish Countryside* (London, Croom Helm).

Donaldson, G. (ed.) (1949) *Accounts of the Collectors of Thirds of Benefices 1561–1572* (Edinburgh, Scottish Historical Society, Third Series, Vol. 42).

Donaldson, G. (ed.) (1970) *Scottish Historical Documents* (Glasgow, Wilson).

Donaldson, J.E. (ed.) (1984) *The Mey Letters* (Sydney, Australia).

Douglas, R. (1934) *Annals of the Royal Burgh of Forres* (Elgin).

Driscoll, S.T. (2002) *Alba: The Gaelic Kingdom of Scotland AD 800–1124* (Edinburgh, Birlinn).

Drummond-Norie, W. (1898) *Loyal Lochaber* (Glasgow, Morison).

Duffy, C. (2007) *The '45* (London, Phoenix).

Duncan, A.A.M. (1978) *Scotland: The Making of the Kingdom* (Edinburgh, Mercat).

Dunlop, J. (1978) *The British Fisheries Society 1786–1893* (Edinburgh, John Donald).

Edwards, K.J. and Ralston, I. (1977–78) 'New Dating and Environmental Evidence from Burghead Fort, Moray', *Proceedings of the Society of Antiquaries of Scotland* 109, 202–210.

Evans, J. (2005) *The Gentleman Usher: The Life and Times of George Dempster 1732–1818* (Barnsley, Pen and Sword).

*Fasti Ecclesiae Scoticanae: The succession of ministers since the Church of Scotland from the Reformation* (1926) ed. H. Scott (Edinburgh, Oliver and Boyd).

Finlay, I. (1979) *Columba* (London, Gollancz).

Fletcher, P. (2010) *Directors, Dilemmas and Debt: The Great North of Scotland and Highland Railways in the Mid-Nineteenth Century* (Aberdeen, Great North of Scotland Railway Association).

Fraser, J. (1905) *Chronicles of the Frasers: The Wardlaw Manuscript*, ed. W. Mackay (Edinburgh, Scottish Historical Society).

Fraser, J.E. (2005) *The Roman Conquest of Scotland: The Battle of Mons Graupius AD 84* (Stroud, Tempus).

Fraser, J.E. (2009) *From Caledonia to Pictland: Scotland to 795* (Edinburgh, Edinburgh University Press).

Gifford, J. (1992) *Highlands and Islands*, The Buildings of Scotland (London, Penguin).

Graham, C. (1977) *Portrait of the Moray Firth* (London, Hale).

Grant, A. (1993) 'The Wolf of Badenoch', in W.D.H. Sellar (ed.) *Moray: Province and People* (Edinburgh, Scottish Society for Northern Studies).

Grant, A. (1996) *Independence and Nationhood: Scotland 1306–1469* (Edinburgh, Edinburgh University Press).

Grant, E. (1972) *Memoirs of a Highland Lady 1797–1827* (London, John Murray).

Gregory, R.A. (2001) 'Excavations by the Late G.D.B. Jones and C.M. Daniels along the Moray Firth Littoral,' *Proceedings of the Society of Antiquaries of Scotland* 131, 177–222.

Haldane, A.R.B. (1973) *The Drove Roads of Scotland*, 3rd edition (Newton Abbot, David & Charles).

Heald, A., Jackson, A. (2001) 'Towards a New Understanding of Iron-Age Caithness', *Proceedings of the Society of Antiquaries of Scotland* 131, 129–147.

*Heimskringla: Sagas of the Norse Kings* (1961) trans. S. Laing, revised P. Foote (London, Dent).

Hume Brown, P. (ed.) (1893) *Scotland before 1700 from Contemporary Documents* (Edinburgh, Douglas).

Hunter, F. (2002) 'Silver for the Barbarians', *Minerva* 13(3), 54–56.

Hunter, F. (2007) *Beyond the Edge of the Empire – Caledonians, Picts and Romans* (Rosemarkie, Groam House Museum).

Hunter, J. (1995) *On the Other Side of Sorrow: Nature and People in the Scottish Highlands* (Edinburgh, Mainstream).

Hustwick, I. (1994) *Moray Firth Ships and Trade during the Nineteenth Century* (Aberdeen, Scottish Cultural Press).

Jackson, A. (1984) *The Symbol Stones of Scotland* (Kirkwall, Orkney Press).

Johnston, P. (1997) 'Older Scots Phonology and Its Regional Variation', in C. Jones (ed.) *The Edinburgh History of the Scots Language* (Edinburgh, Edinburgh University Press).

Kiedel, K.P. and Schnall, U. (eds) (1985) *The Hanse Cog of 1380* (Bremerhaven, Förderverein Deutsches Schiffarhtsmuseum).

Konstam, A. (2010) *Strongholds of the Picts: The Fortifications of Dark Age Scotland* (Oxford, Osprey).

Kyd, J.G. (1952) *Scottish Population Statistics* (Edinburgh, Scottish Historical Society, Third Series, Vol. 49).

Laing, L. (2000) 'How Late Were Pictish Symbols Employed?', *Proceedings of the Society of Antiquaries of Scotland* 130, 637–650.

Laing, L. and Laing, J. (1994) *The Picts and the Scots* (Stroud, Sutton).

Lamb, H.H. (1977) *Climate: Past, Present and Future* (London, Methuen).

Larner, C. (1981) *Enemies of God: The Witch Hunt in Scotland* (London, Chatto & Windus).

Lauder, T.D. (1998 [1873]) *The Great Moray Floods of 1829*, new edition (Elgin, Moray Books).

Lawrie, Sir Archibald (1905) *Early Scottish Charters Prior to AD 1153* (Glasgow, James MacLehose and Sons).

Lippe, R. (ed.) (1890) *Selections from Woodrow's Biographical Collections: Divines of the North-east of Scotland* (Aberdeen, New Spalding Club).

Livingstone, A., Aikman, C.W.H. and Hart, B.S. (2001) *No Quarter Given: The Muster Roll of Prince Charles Edward Stuart's Army 1745–46* (Glasgow, Wilson).

Lynch, M. (1991) *Scotland: A New History* (London, Pimlico).

Lynch, M. (ed.) (2001) *The Oxford Companion to Scottish History* (Oxford, Oxford University Press).

McAndrew, B.A. (1999) 'The Sigillography of the Ragman Roll', *Proceedings of the Society of Antiquaries of Scotland* 129, 663–752.

Macleod, D. (1892) *Gloomy Memories in the Highlands of Scotland*, facsimile edition (Fort William, Nevisprint).

Macdougall, N. (1989) *James IV* (Edinburgh, John Donald).

MacGill, W. (ed.) (1909) *Old Ross-shire and Scotland as seen in the Tain and Balnagown Documents* (Inverness).

McGrail, S. (2006) *Ancient Boats and Ships* (Princes Risborough, Shire).

McIntyre, A. (1998) 'Survey and Excavation at Kilearnan Hill, Sutherland, 1982–3', *Proceedings of the Society of Antiquaries of Scotland* 128, 167–201.

Mackay, W. and Boyd, H.C. (eds) (1911) *Records of Inverness: Vol. I: Burgh Court Books 1556–86* (Aberdeen).

Mackay, W. and Laing, G.S. (eds) (1924) *Records of Inverness: Vol. II: 1603–1688* (Aberdeen).

Mackenzie, A. (1891) *History of the Chisholms* (Inverness).

Mackenzie, A. (1894) *History of the Mackenzies* (Inverness).

Mackenzie, A. (1898) *History of the Munros of Fowlis* (Inverness).

Mackintosh, A.M. (1903) *The Mackintoshes and Clan Chattan* (Edinburgh).

McKean, C. (1987) *The District of Moray: An Illustrated Architectural Guide* (Edinburgh, Scottish Academic Press/RIAS).

McKean, C. (1990) *Banff and Buchan: An Illustrated Architectural Guide* (Edinburgh, Mainstream/RIAS).

McKirdy, A., Gordon, J. and Crofts, R. (2007) *Land of Mountain and Flood: The Geology and Landforms of Scotland* (Edinburgh, Birlinn).

McNeill, P. and Nicholson, R. (1975) *An Historical Atlas of Scotland c.400–c.1600* (St Andrews, Conference of Scottish Medievalists).

Martin, M. (1994) *A Description of the Western Isles of Scotland circa 1695*, ed. D.J. Macleod (Edinburgh, Birlinn).

Mason, D.J.P. (2003) *Roman Britain and the Roman Navy* (Stroud, Tempus).

Maxwell, G.S. (1989) *The Romans in Scotland* (Edinburgh, Mercat Press).

Millar, A.H. (ed.) (1909) *A Selection of Scottish Forfeited Estates Papers* (Edinburgh, Scottish Historical Society).

Miller, H. (1869 [1834]) *Scenes and Legends of the North of Scotland*, 7th edition (Edinburgh, Nimmo).

Miller, J. (1994) *A Wild and Open Sea: The Story of the Pentland Firth* (Kirkwall, Orkney Press).

Miller, J. (1999) *Salt in the Blood* (Edinburgh, Canongate).

Miller, J. (2001) *A Caithness Wordbook* (Wick, North of Scotland Newspapers).

Miller, J. (2004) *Inverness* (Edinburgh, Birlinn).

Miller, J. (2007) *Swords for Hire: The Scottish Mercenary* (Edinburgh, Birlinn).

Millward, R. (1964) *Scandinavian Lands* (London, Macmillan).

Mitchell, A. (ed.) (1902) *Inverness Kirk Session Records 1661–1800* (Inverness).

Mitchell, A. (ed.) (1906) *Geographical Collections Relating to Scotland Made by Walter MacFarlane* (Edinburgh).

Mitchell, J. (1884) *Reminiscences of My Life in the Highlands* (Inverness).

Moffat, A. and Wilson, J. (2011) *The Scots: A Genetic Journey* (Edinburgh, Birlinn).

Mowat, I.R.M. (2003) *Easter Ross 1750–1850: The Double Frontier* (Edinburgh, Birlinn).

Munro, J. (1984) 'The Clan Period', in D. Omand (ed.) *The Ross and Cromarty Book* (Golspie, The Northern Times Ltd).

Nicolaisen, W.F.H. (1976) *Scottish Place-Names* (London, Batsford).

Nicolaisen, W.F.H. (1993) 'Names in the landscape of the Moray Firth', in W.D.H. Sellar (ed.) *Moray: Province and People* (Edinburgh: Scottish Society for Northern Studies).

Oppenheimer, S. (2007) *The Origins of the British* (London, Robinson).

*Orkneyinga Saga* (1981) trans. H. Palsson and P. Edwards (London, Penguin).

Owen, J.S. (1995) *Coal Mining at Brora* (Inverness, Highland Libraries).

Paton, H. (ed.) (1895) *The Lyon in Mourning*, 3 vols (Edinburgh, Scottish History Society).

Paton, H. (1903) *The Mackintosh Muniments* (Edinburgh).

Pottinger, M. (1997) *Parish Life on the Pentland Firth* (Thurso, Whitemaa).

Prebble, J. (1962) *Culloden* (London, Atheneum).

*Registrum Episcopatus Moraviensis* (1837) (Edinburgh, Bannatyne Club).

Reynolds, P.J. (1995) 'Rural Life and Farming' in M. Green (ed.) *The Celtic World* (London, Routledge) (reprinted in Butser Ancient Farm Occasional Papers, III).

Richards, E. (1973) *The Leviathan of Wealth: The Sutherland Fortune in the Industrial Revolution* (London, Routledge).

Richards, E. (2008 [1982]) *The Highland Clearances* (Edinburgh, Birlinn) (originally published as *A History of the Highland Clearances*, London, Croom Helm).

Ritchie, A. (1976) 'Excavation of Pictish and Viking-Age Farmsteads at Buckquoy, Orkney', *Proceedings of the Society of Antiquaries of Scotland* 108, 174–227.

Ritchie, A. (1989) *Picts* (Edinburgh, HMSO).

Robertson, B. (2011) *Lordship and Power in the North of Scotland: The Noble House of Huntly 1603–1690* (Edinburgh, John Donald).

Robertson, J. (ed.) (1843) *Collections for a History of the Shires of Aberdeen and Banff*, (Aberdeen, Spalding Club).

Ross, D. (2001) *Scottish Place-names* (Edinburgh, Birlinn).

Ross, J.R. (1972) *The Great Clan Ross* (Canada, Deyell).

Royle, T. (2007) *The Flowers of the Forest: Scotland and the First World War* (Edinburgh, Birlinn).

Royle, T. (2011) *A Time of Tyrants: Scotland and the Second World War* (Edinburgh, Birlinn).

Sage, D. (1899) *Memorabilia Domestica: Parish Life in the North of Scotland* (Wick).

Salter, M. (1995) *The Castles of Western and Northern Scotland* (Malvern, Folly Publications).

Schildhauer, J. (1985) *The Hansa: History and Culture* (Leipzig, Edition Leipzig).

Sellar, W.D.H. (ed.) (1993) *Moray: Province and People* (Edinburgh, Scottish Society for Northern Studies).

Shaw, L. (1882) *The History of the Province of Moray* (Glasgow).

Sissons, J.B. (1967) *The Evolution of Scotland's Scenery* (Edinburgh, Oliver and Boyd).

Small, A. (1983) 'Dark Age Scotland', in G. Whittington and I.D. Whyte (eds) *An Historical Geography of Scotland* (London, Academic Press).

Smith, A.P. (1984) *Warlords and Holy Men: Scotland AD 80–1000* (London, Arnold).

Smith, C. (2000) 'A Grumphie in the Sty: An Archaeological View of Pigs in Scotland, from Their Earliest Domestication to the Agricultural Revolution', *Proceedings of the Society of Antiquaries of Scotland* 130, 705–724.

Smout, T.C. (1969) *A History of the Scottish People 1560–1830* (Glasgow, Collins).

Smout, T.C. (1986) *A Century of the Scottish People 1830–1950* (Glasgow, Collins).

Spalding, J. (1850) *Memorialls of the Trubles in Scotland and in England*, 2 vols (Aberdeen, Spalding Club).

Stone, J. (1991) *Illustrated Maps of Scotland from Blaeu's Atlas Novus of the 17th Century* (London, Studio Editions).

Strang, A. (1998) 'Recreating a Possible Flavian Map of Roman Britain with a Detailed Map for Scotland', *Proceedings of the Society of Antiquaries of Scotland* 128, 425–440.

Sutherland, I. (1984) *Wick Harbour and the Herring Fishing* (Wick).

Sutherland, I. (1985) *From Herring to Seine Net Fishing on the East Coast of Scotland* (Wick).

Symon, J.A. (1959) *Scottish Farming: Past and Present* (Edinburgh, Oliver and Boyd).

Vallance, H.A. (1996) *The Highland Railway* (Colonsay, Lochar).

Tacitus (1970) *The Agricola and the Germania*, trans. H. Mattingly, revised S.A. Handford (London, Penguin).

Tayler, A. and Tayler, H. (1914) *The Book of the Duffs* (Edinburgh).

Tayler, A. and Tayler, H. (1937) *The House of Forbes* (Aberdeen, Third Spalding Club).

Taylor, A.B. (1980) *Alexander Lindsay: A Rutter of the Scottish Seas c.1540* (Greenwich, National Maritime Museum, Monographs and Reports 44).

Taylor, S. (2002) *Place-Name Survey of the Parishes of Kilmorack, Kiltarlity & Convinth, and Kirkhill, Inverness-shire.* Available online at http://arts.st-andrews.ac.uk/beauly.

Thomas, D. (ed.) (1977) *Wales: A New Study* (Newton Abbot, David and Charles).

Thomson, D.S. (ed.) (1983) *The Companion to Gaelic Scotland* (Oxford, Blackwell).

Turner, J. (1986) *Scotland's North Sea Gateway: Aberdeen Harbour AD 1136–1986* (Aberdeen, Aberdeen University Press).

Watson, W.J. (1904) *Place-Names of Ross and Cromarty* (Inverness).

Watson, W.J. (1993 [1926]) *The Celtic Place-names of Scotland* (Edinburgh, Birlinn).

Watt, D. (2007) *The Price of Scotland: Darien, Union and the Wealth of Nations* (Edinburgh, Luath).

Waugh, D. (1989) 'Place-names', in D. Omand (ed.) *The New Caithness Book* (Wick, North of Scotland Newspapers).

Whittington, G. (1974–75) 'Placenames and the Settlement Pattern of Dark-Age Scotland', *Proceedings of the Society of Antiquaries of Scotland* 106, 99–110.

Whittington, G. and Whyte, I.D. (1983) *An Historical Geography of Scotland* (London, Academic Press).

Wilson, R.J.A. (1980) *A Guide to the Roman Remains in Britain* (London, Constable).

Woolf, A. (2006) 'Dun Nechtain, Fortriu and the Geography of the Picts', *Scottish Historical Review* 85(2), 182–201.

Woolf, A. (2007) *From Pictland to Alba: 789–1070* (Edinburgh, Edinburgh University Press).

Wordsworth, J. (1999) 'A Later Prehistoric Settlement at Balloan Park, Inverness', *Proceedings of the Society of Antiquaries of Scotland* 129, 239–249.

Worthington, D. (2011) 'A Northern Scottish Maritime Region: The Moray Firth in the Seventeenth Century', *International Journal of Maritime History* 23(2).

Young, A. (1997) *Robert the Bruce's Rivals: The Comyns 1212–1314* (East Linton, Tuckwell).

# PICTURE ACKNOWLEDGEMENTS

The colour photographs are reproduced with kind permission of the following:

Jim Henderson; Moray Firth Partnership/by Scotavia Images; National Museums Scotland; Trustees of the National Gallery of Scotland.

The black-and-white photographs and line illustrations are reproduced with kind permission of the following:

Diocese of Moray, Ross and Caithness of the Scottish Episcopal Church, and Highland Libraries/Am Baile: pages 126, 143; J. Gayton: page 21; Highland Council, licensed to High Life Highland: page 195; Highland Libraries/Am Baile: pages 95, 137, 139, 140, 143, 158, 160, 198; Avril Hill: page 128; The Moray Firth Partnership/by Scotavia Images: pages 5, 6, 8, 10, 11, 12, 33, 46, 138, 153, 157, 176, 185, 192, 209, 214, 216, 220; Charlie Phillips Images: page 223 (dolphins); Scran/Dianne Sutherland: page 187; Scran/Moray Council: page 200; Scran/National Museums Scotland: pages 22, 23; Scran/RCAHMS: page 32; Trustees of the Conference of Scottish Medievalists: page 63, from p.127 of McNeill and Nicholson 1975, original map and text by Richard Muir; page 64, from p.128 of McNeill and Nicholson 1975, original map and text by Geofrey Stell; page 65, from p.129 of McNeill and Nicholson 1975, original map and text by Richard Muir; Dr Doreen Waugh and North of Scotland Newspapers Ltd: page 48; The Wick Society, Johnston Collection: pages 181, 196.

# INDEX